RIVER MEDWAY
PLEASURE STEAMERS

The coal-fired *Kingswear Castle* on the River Medway at the end of her restoration. She is now the Medway's last operational paddle steamer.

RIVER MEDWAY
PLEASURE STEAMERS

ANDREW GLADWELL

AMBERLEY

Medway Queen departing from Strood Pier in 1963.

To find out more about paddle steamer heritage: www.heritagesteamers.co.uk

To book a cruise on the River Medway's paddle steamer *Kingswear Castle*:
www.kingswearcastle.co.uk

To book a cruise on *Waverley* and *Balmoral*: www.waverleyexcursions.co.uk

To join the PSPS: www.psps.freeserve.co.uk

To find out more about *Medway Queen*: www.medwayqueen.co.uk

To find more books by Andrew Gladwell: www.andrewgladwell.co.uk

First published 2010

Amberley Publishing Plc
Cirencester Road, Chalford,
Stroud, Gloucestershire, GL6 8PE

www.amberley-books.com

Copyright © Andrew Gladwell 2010

The right of Andrew Gladwell to be identified as the
Author of this work has been asserted in accordance
with the Copyrights, Designs and Patents Act 1988.

ISBN 978 1 84868 695 3

British Library Cataloguing in Publication Data.
A catalogue record for this book is available from the
British Library.

Typeset in 10pt on 12pt Sabon.
Typesetting and Origination by FONTHILLDESIGN.
Printed in the UK.

Contents

Acknowledgements

In compiling this book, I have appreciated the generosity of the individuals below, who have given so much of their time, material and so many of their memories to ensure that the heritage of the often forgotten River Medway pleasure steamers is preserved. The story of the Medway pleasure steamers as well as those from the rest of the UK is preserved in the Paddle Steamer Preservation Society Collection (www.heritagesteamers. co.uk). The PSPS Collection is the largest and most important collection of paddle and pleasure steamer material in the UK. It provides a permanent and accessible home for unique material such as photographs, handbills, models, fittings and posters, which are so intertwined with the memories included in this book. I would like to thank Campbell McCutcheon, Kieran McCarthy and Alan Peake for their assistance with this book. Alan has a particular link with the Medway fleet in that his late father, William Peake, was Managing Director of the New Medway Steam Packet Company towards the end of its existence. In particular, I would like to say a very special thank-you to John Richardson, MBE, and Roy Asher for their great enthusiasm and generosity in allowing me to use their unique material, which captures so vividly the great days of the well-loved 'Queen Line Steamers' of the River Medway.

Introduction

Many people remember with great fondness the pleasure steamers that plied the River Thames in the years after the end of the Second World War. The mighty General Steam Navigation Company, more commonly known as 'Eagle Steamers', dominated the business. But fewer people now appreciate the significant role that the New Medway Steam Packet Company (known as 'Queen Line Steamers') played in the growth of services in the Thames Estuary from the early 1920s until the early 1960s. The 'Eagle & Queen Line of Steamers' became one of the most formidable pleasure steamer operators in the whole of the UK before its untimely demise in the 1960s.

The roots go back to 1837 when the Medway Steam Packet Company was formed to provide a steamer service between Rochester and Sheerness. Services were soon extended to connect with the burgeoning and lively Essex resort of Southend-on-Sea thereby forming the most useful and profitable of routes, which would become beloved of daytrippers until regular services from the Medway ceased in September 1963.

With several restructures, the Medway services survived amidst a scene of moderate prosperity throughout the latter part of the nineteenth century and early part of the twentieth century. Famous steamers such as *City of Rochester*, *Princess of Wales* and *Lady of Lorne* provided a reliable and well-loved fleet of steamers. By the end of the First World War, things were changing, and with that curious blend of personalities, ships, a bold vision and historical events, pleasure steamer services on the River Medway were to have their finest hour and the residents of the Medway towns were about to sample the most-loved paddle steamers that ever plied the river.

The catalyst for change was Captain Sydney Shippick. By the early 1920s, he set about changing Medway services out of all recognition from that of a generation earlier. The New Medway Steam Packet Company was formed in December 1919 from the older company with the prime aim of reinvigorating services. Captain Sydney Shippick became an enigmatic director of the company. The older paddle steamers *City of Rochester* and *Princess of Wales* were joined by the diminutive *Audrey*. Shippick soon became a major player in the company and was appointed Managing Director in 1923. He, along with Captain Tommy Aldis, then embarked upon a massive and ambitious enlargement of the company. One by one, new steamers were acquired, and in 1924, perhaps the most famous, well-loved and longest-lived paddle steamer entered service – the *Medway Queen*. Built by Ailsa of Troon, *Medway Queen* was placed on a service from Chatham and Southend to Clacton, Walton and Felixstowe. Under the helm of Sydney Shippick and Tommy Aldis, steamer services eventually extended as far as Great Yarmouth and Ramsgate and across the English Channel to Boulogne and Calais.

By the 1930s, the Queen Line of Steamers was growing out of all recognition from that of a decade earlier. Steamers such as *Queen of Kent*, *Queen of Thanet*, *Essex Queen*, *Queen of the South*, and *Queen of Southend* were often modernised and placed on new routes, which expanded the services of the Medway Company. These new routes attacked the previously strong dominance of the mighty General Steam Navigation Company. The spirit of competition brought with it more defined services, an eagerness to upgrade vessels, as well as often causing confusion to passengers provided with a plethora of services at Thames piers – where would it all end?

In 1935, perhaps the biggest revolution came when the sleek, modernistic and simply wonderful motor vessel *Queen of the Channel* entered service on London's river. This was perhaps the pinnacle of Sydney Shippick's dream. Now, a large, economical and modern new-look vessel could provide efficient and streamlined steamer services in an age immortalised in the Art Deco style. With the introduction of *Royal Sovereign* in 1937 and the ultimate pleasure steamer *Royal Daffodil* in 1939, the recently amalgamated New Medway Steam Packet Company and General Steam Navigation provided the ultimate in pleasure steamer services on the London River. Surely, nothing would now affect the mighty new Eagle & Queen Line of Steamers on the Medway and Thames?

The Second World War affected pleasure steamers on the River Medway as much as it did every other aspect of life in the UK. Many steamers were lost and many more faced an uncertain and short career in the austere postwar years as Britain attempted to gain some degree of normality after six years of war. The Second World War was also a time of great heroism for the Medway paddle steamers as the charming peacetime steamer *Medway Queen* became a heroine of Dunkirk. At the end of the conflict, both *Royal Sovereign* and *Queen of the Channel* had become wartime losses. Quickly, shiny and equally resplendent new vessels were built to replace them. But the confidence shown by these two vessels didn't quite match the new way of the world. Yes, people wanted to escape from the suppression of normal life caused by war, but they also wanted a new freedom. The motor car and foreign holiday with their flexibility and non-reliance on weather caused a gradual decline in those opting for a fortnight at the seaside by pleasure steamer. Services valiantly continued until 1963 by the queen of River Medway paddle steamers – the *Medway Queen*. In September of that year, she made her final farewell to the resorts of Herne Bay and Southend, thereby closing the final chapter of paddle steamer services on the River Medway.

But the tearful farewell of the *Medway Queen* in 1963 wasn't the end. The Thames and Medway later witnessed the withdrawal of the three mighty motor vessels in the mid-1960s, as well as brief visits by steamers such as *Queen of the South* (ex-*Jeanie Deans*) and *Consul*. During the early 1970s, a paddle steamer quietly entered the River Medway for the first time – she was the *Kingswear Castle*. She was arriving on the Medway for a lengthy restoration. Few would have realised at the time that *Kingswear Castle* would ultimately have a career on the Medway that would be longer than most of her predecessors. This is even more remarkable as it happened at a time when those who remembered a pleasure steamer cruise were dwindling. Over the years, *Kingswear Castle* has witnessed many other pleasure steamer visitors to the River Medway, both big and small. These have ranged from *Balmoral*, *Clyde*, *Princess Pocahontas* to the world-famous *Waverley*. Another paddle steamer built in 1924, the famous *Medway Queen*, also returned during the 1980s and is at present undergoing a restoration project to return her to her 1924 glory. The story of pleasure steamer services on the River Medway is therefore still evolving, and now, almost two centuries since services started, it looks certain that the story will continue for many more years to come!

Chapter 1

Early Years

Passengers enjoying a cruise aboard the popular and well-loved *City of Rochester* in the days before the First World War.

The first recorded paddle steamer services on the River Medway commenced in 1837 when the Medway Steam Packet Company was founded. This original company was formed with the main aim of linking Chatham and Sheerness. This was a distance of around nine miles by water in the days before a reliable and fast railway service was available. The earliest boats were built of mahogany and were repaired and kept in good condition for service in the company's own shipyard at Rochester.

Little is known of the earliest steamers. The first *City of Rochester* was built in 1849 and was initially privately owned. Some twenty-two years later, she was acquired by the Medway Steam Packet Company, who operated her between Rochester, Chatham, Sheerness and Southend. Little is also known also of *Alma*. She was built in 1855 and had a long career before being broken up in 1899. Like the first *City of Rochester*, she had once been owned by Mr Giles from the Medway area.

Towards the end of the nineteenth century, in 1888, the Medway Company acquired the Greenock-built *Lady Margaret* from the Bristol Channel Express Company. Unlike the other steamers, which were built of wood, she was built of iron. She was also around twenty feet longer than the other steamers in the fleet. Her service came to an end in 1903 when she was destroyed by fire. *Lady Margaret* was a great asset to the aging fleet. By the late 1890s, these steamers met their end either by being broken up or by being destroyed by accident. *City of Rochester* remained in service until being broken up in 1897 after serving Rochester, Chatham, Sheerness and Southend for almost fifty years.

At the turn of the century, two of the most famous and well-loved Medway Steam Packet steamers were entering service. These were the *Princess of Wales* (1896) and *City of Rochester* (1904). They entered service in the halcyon days before the First World War and carried countless thousands of passengers between the Medway towns, Sheerness and Southend, as well as further, until hostilities broke out in 1914.

The First World War saw the Medway paddle steamers called up for war service in 1915. They were joined by the small and important *Audrey*. She entered the Medway paddle steamer stage to work as a ferry. When services recommenced in 1919, it was the start of an era of expansion and change. These were changes that would see the Medway Steam Packet Company change out of all recognition from that of a generation earlier.

MEDWAY STEAM PACKET COMPANY.

Sir,

I am instructed by the Committee to state that they have been enabled to allot you _____ Shares in the above Company, and you are requested on or before Friday next, the 8th instant, to pay into the Chatham Bank the sum of £ _____ being a Deposit of Ten Shillings on each Share.

A Meeting of all Shareholders who shall have paid the above Deposit will be held at the Sun Inn, Chatham, on Friday next, at Seven o'Clock in the Evening, to choose the Directors, Auditors, Trustees and Treasurer, for the Chatham Division.

I am, Sir,

Your obedient Servant,

Walter Hills, Secretary.

Chatham,

2nd July, 1836.

P.S. You will please to observe that no Shareholder who has not paid his Deposit by Friday next, will be entitled to vote.

Document confirming the issue of shares in the Medway Steam Packet Company dated 2 July 1836 at the inception of the company. The document was issued by Walter Hills, the secretary of the company. Shareholders were asked to pay a deposit of 10s for each share issued to the Chatham Bank by the following Friday. The first meeting of shareholders was held at the Sun Inn at Chatham on that date.

Lady of Lorne on the River Medway around 1890. The Medway Steam Packet Company was formed in 1837 with the main aim of linking Chatham with Sheerness. Information on the steamers prior to the acquisition of *Lady of Lorne* is scarce. The earliest paddle steamers on the Medway were built of mahogany.

Lady of Lorne departing from the old wooden pier at Southend around 1890. She had been built at Rochester in 1871 and was over 116 feet long. She had a relatively short career on the Medway until 1899.

A view looking towards Strood and its pier from Rochester Castle. Strood Pier was built in 1905 at a cost of £1,520. Thereafter, cruises were available from there as well as from Chatham. A paddle steamer can be seen embarking passengers at the pier.

Postcard sent by a passenger aboard *Princess of Wales* in June 1910. The view shows Frindsbury in the background. Note the awning at the stern of the vessel. *Princess of Wales* was built in 1896 in Middlesbrough and was the first Medway steamer to run the regular service between Rochester and Southend. During the nineteenth century, Medway services had mainly been confined to those between Chatham and Sheerness.

THE MEDWAY FROM CHATHAM PIER

A view showing *City of Rochester* and two other paddle steamers from the Medway fleet moored midway between Rochester and Chatham close to Sun Pier.

Pleasure Steamer "Princess of Wales" Chatham.

Princess of Wales on the River Medway at Chatham around 1905. She continued on Medway service until being chartered by the Stanley-Butler Steamship Company of Kirkcaldy for further service on the Firth of Forth. *Princess of Wales* unfortunately sank at her moorings in March 1927. She was repaired and returned to the River Medway before ultimately being sold for scrap to T. Wards of Grays in October 1928.

Sheerness Pier was a frequently used calling point for New Medway Steam Packet paddle steamers. It was built in the nineteenth century at the same time as a promenade was built at Sheerness, making it a popular resort. The access to the pier was in the Bluetown area.

Sheerness Pier around 1919. The pier was crucial to Medway services, especially in the early days before services to Southend and further afield developed.

Passengers on the promenade deck of *City of Rochester* on 28 July 1911. The photograph was taken by local photographer G. Morris of High Street, Rochester. In the days before the advent of mass camera use, passengers would pose in such a photograph and would then collect their precious photograph on the return journey. For many, they would only manage one day at the seaside a year and therefore wore their best clothes. A souvenir photograph was a necessity and may well have been the only photograph taken of them during each year.

The New City of Rochester approaching Chatham Pier.

City of Rochester approaching Chatham Sun Pier around 1905. *City of Rochester* was built by Scott of Kingholm in 1904. She was the largest steamer in the fleet at the time of her entry into service. She was 160.2 feet in length had a breadth of 22.2 feet and a depth of 7.3 feet. This image shows the pier with its original large, square-shaped shelter. In the distance behind the steamer are the Royal Marine Barracks with the adjacent St Mary's church. The church housed the Medway Heritage Centre during the 1980s, which told the story of the River Medway and its towns.

City of Rochester at sea. *City of Rochester* and *Princess of Wales* operated services from the Medway at the start of the twentieth century. They offered two daily trips, including Fridays from Strood (for Rochester) calling at Chatham, Gillingham and Sheerness. They then crossed to Southend, arriving around 11.15 a.m. with an immediate return. A second trip was made during the afternoon.

A fine view of *City of Rochester* during her heyday with her decks well packed with passengers. In 1922, *City of Rochester* re-entered service under the command of Captain Tommy Aldis. *City of Rochester* had a service speed of around 14 knots, whilst *Princess of Wales* only managed 11 knots.

A view of passengers aboard *City of Rochester* on 28 July 1911. Note the details on the deck, including the sponson, gangway and ventilators. Passengers shown in this image would face many changes during the First World War. The Medway Steam Packet fleet also witnessed a period of change when the happy peacetime paddle steamers went to war. The River Medway that they returned to would be a very different place due to the emergence of one man – Captain Sydney Shippick.

A view of *City of Rochester* cruising at speed with the empty fields of Frindsbury in the background around 1907. The steamer is looking particularly fresh and immaculate, which helps to date the photograph close to her entry into service on the River Medway in 1904.

A postcard showing *City of Rochester* cruising in the Medway and showing four scenes to welcome visitors to Sheerness. At the time, Sheppey was a popular magnet for tourists with its promenade. Steamer passengers would have arrived at the pier head shown in the top right-hand corner of the card.

Passengers aboard *City of Rochester* on 24 September 1911. It shows a crowded promenade deck and note that everyone is wearing a hat. The baby in the centre of the photograph looks particularly well wrapped up in many layers of clothing. *City of Rochester* was the second paddle steamer on the Medway to bear that name. The first was built in the late 1840s and was acquired by the Medway Company during the 1870s before being scrapped in 1897. *City of Rochester*, shown in this photograph, entered service in 1904.

Chatham Dockyard and Pleasure Steamer "Princess of Wales", which runs between Strood, Chatham and Southend during the Season.

Princess of Wales passing Chatham Dockyard around 1900. *Princess of Wales* was built for the Medway Company by Craggs & Son at Middlesbrough. Initially, she was used on the Rochester and Southend service so would have passed this scene almost every day. The dockyard scene in the distance shows the Victorian extension on St Mary's Island, where three gigantic basins were built. The building with the tall chimney just behind the funnel of *Princess of Wales* is the Pumping House. It now survives as one of the few remaining buildings of the Victorian extension. For several years, *Medway Queen* was moored in front of this building awaiting restoration before being moved to Damhead Creek in the late 1980s.

City of Rochester on the River Medway around 1907. *City of Rochester* was the last paddle steamer acquired by the Medway Steam Packet Company before the company was re-formed in 1919.

A tranquil image of *Princess of Wales* on the River Medway around 1905.

City of Rochester alongside Upnor Pier around 1910. Upnor Pier was a popular calling point for Medway paddle steamers.

Chapter 2

A Brave New Vision

The years between the two world wars witnessed a great expansion of paddle steamer services on the River Medway, pioneered by Captain Shippick.

The decade after the end of the First World War witnessed the greatest ever expansion of Medway pleasure steamer services. Central to this story is one man and his ship: Captain Sydney Shippick and *Audrey*.

In December 1919, the Medway Steam Packet Company was re-formed as the New Medway Steam Packet Company with Captain Shippick being made a director. Shippick had originally travelled to the Medway and had run his small steamer *Audrey* on charter to the Admiralty. In 1915, she was purchased by the Admiralty and used at Chatham on ferry services.

The Medway Steam Packet Company announced on the 16 June 1921 that it would thereafter be known as the New Medway Steam Packet Company. After this change in name, it built up its fleet equal to that of the Belle Steamers. In 1922, *Audrey* was re-purchased by Shippick for the New Medway Company, who then reconditioned her. This gesture was the beginning of an era of great expansion for the company. When appointed Managing Director in 1923, Sydney Shippick introduced Captain Tommy Aldis to the company, and the duo immediately set about opening up exciting new routes and services. They were assisted to a great extent by the New Medway Company's extensive shipyard at Rochester, where they could maintain their steamers as well as offering employment during the winter months. *Audrey* had the distinction of opening the New Medway Company's regular service to Herne Bay. The inaugural cruise was under the command of Captain Tommy Aldis on Sunday 20 May 1923.

Within a short period of time, services were offered to resorts as far away as Great Yarmouth and Ramsgate, as well as to Boulogne and Calais. By 1930, the New Medway Steam Packet Company was a serious competitor to the General Steam Navigation Company.

Board of Trade regulations after the 1914-18 war had became more stringent for crossing the English Channel. None of the New Medway Company paddle steamers were able to comply with these regulations, so they purchased two Admiralty Ascot-class minesweepers that were about to go to the breakers. They were named *Atherstone* and *Melton*. *Queen of Kent* was acquired by Shippick because he saw her and her sister as the most cost-effective way of expanding the fleet quickly. They were converted in the company's shipyard to become *Queen of Kent* and *Queen of Thanet*. Day trips to the Continent were a revolution in the 1930s. Their introduction had opened up a significant market for operators. But, this revolution needed a new, faster and more comfortable style of pleasure steamer to drive the revolution forward. In 1935, the new diesel motor ship *Queen of the Channel* was designed and built with similar proportions to a paddle steamer. It was a joint project between the New Medway Company and Denny of Dumbarton. A new company called the London & Southend Continental Steamship Company was formed for this purpose. It was, of course, very profitable, and the New Medway Company were able to pay off the remaining money owed to Denny's with a substantial loan from Lloyds Bank. The London & Southend Continental Steamship Company was wound up and the ship became a full member of the New Medway Company fleet. Sydney Shippick then felt eager to expand the fleet further. The New Medway Company placed a new contract for a second Denny-built motor ship in October 1936 due to the success of *Queen of the Channel*. *Queen of the Channel* was mortgaged to pay for the new Denny-built motor ship *Royal Sovereign*. Thus, the new Eagle & Queen Line found themselves in a healthy position in 1939. It would appear that the company had grand plans to further expand services around the coastline, past Ramsgate along the South Coast. Alas, these came to nothing, as the Second World War was declared soon after.

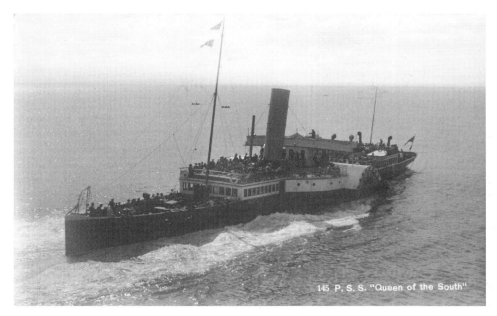

1924 witnessed two additions to the New Medway Company fleet. *Woolwich Belle* of the Belle Steamer fleet was acquired from Channel Excursion Steamers and was renamed *Queen of the South*. She was delivered at the end of the season, so didn't enter service until 1925. The now-famous *Medway Queen* was the other steamer.

Queen of the South on the grid at Rochester. Her career with the New Medway Steam Packet was short. She was laid up by 1931 and then sold for £300 to be broken up at Grays.

Queen of the South and *Essex Queen* laid up in the River Medway during the mid-1930s. *Essex Queen* was purchased in 1925 and was formerly known as *Walton Belle*. She could carry up to 1,200 passengers. She undertook extensive service along the East Coast at such places as Great Yarmouth, Walton and Clacton.

Essex Queen undertook a charter on Wednesday 30 May 1928 when she landed 334 passengers on a cruise organised by the Shepherds Friendly Society from Clacton to Great Yarmouth. An amusing verse was composed to record the event:

The passengers boarded like lambs, saying – bar, bar, bar
Everybody enjoyed themselves as there were no crooks on board
to fleece the passengers, so nobody lost their wool.
The elderly Rams and Ewes provided mutton for lunch.
Many returned to Clacton red faced, having over enjoyed their dip.

Eagle and Queen Line Pleasure Steamers

ESSEX QUEEN

(or Queen Steamer) leaves

RAMSGATE HARBOUR

(Weather and other circumstances permitting)

SUNDAYS TO THURSDAYS
At 4-30 p.m.

— FOR —

MARGATE, HERNE BAY, SOUTHEND, SHEERNESS and CHATHAM

FARES :

		Single	Return Boat & Coach
To MARGATE	-	1/6	2/-
		Single	Season Return
HERNE BAY	-	2/-	3/6
SOUTHEND	-	4/-	6/-
SHEERNESS	-	3/6	6/-
CHATHAM	-	4/-	6/6

Children under 14 years half-fare.

Free Admission to Pier.

Luggage accompanying passengers up to 100-lb. free.

Passengers are only carried on the terms and conditions printed on the Company's Tickets.

PHONE RAMSGATE 1056

Excellent Refreshments on Board.   Fully Licensed.

Printers: *Bligh and Co., Ltd. Ramsgate.*

Handbill advertising services by *Essex Queen* from Ramsgate. She had been bought by the New Medway Company during its period of rapid expansion in December 1925. *Essex Queen* was one of the longest surviving of Thames and Medway paddle steamers and was built by Denny in 1897.

Essex Queen in her first career as *Walton Belle* saw service during the First World War. She went to White Sea in the Arctic during the Russian campaign – a distance of over 2,000 miles. After the Second World War, she finally appeared as the *Pride of Devon* at Torquay before being scrapped after fifty-four years of service.

Essex Queen can be seen making her approach to Southend Pier in 1929. The Eastern Berthing Arm, more commonly known as the 'Prince George Extension', was built in that year to accommodate the ever-increasing number of paddle steamers visiting Southend. This extension was opened by HRH the Duke of Kent on 8 July 1929. The total length of the steamer berths on the south side of the pier was 540 feet. It was sufficient to accommodate two large pleasure steamers and four small ones. Just before this handbill was published in 1933, 1.25 million people visited the pier. By its heyday in 1949, there were an astounding 7 million visitors a year.

WINGENT'S FLOUR MILLS, CHATHAM SUN PIER.

A wonderful aerial view of Chatham Sun Pier with *Audrey* alongside it during the mid-1920s. The mills to the right of the image are the vast flour mills of Wingent & Kimmins Ltd. The reverse of the postcard shows an advertisement for the company, promoting its fine location on the Medway, the adjacent wharves, as well as the plentiful supply of local wheat. The area is now occupied by a large retail unit, offices and a vast car park. The main change to Sun Pier is that the large, square shelter has now gone.

Audrey at Ramsgate in 1929. *Audrey* went into service for the New Medway Steam Packet Company around Whitsun 1923 and was initially placed on the Sheerness to Southend run. The well-respected Captain Tommy Aldis took command of *Audrey* at the beginning of her Medway career and started the Herne Bay run. *Audrey* worked from Ramsgate towards the end of her career.

Audrey had been owned by Sydney Shippick, who operated her from Bournemouth to Poole, Swanage and Studland. During the early days of the First World War, she moved to the River Medway, and Shippick operated her for the Admiralty as a naval ferry, taking workers to the airship works at Kingsnorth. The Admiralty purchased *Audrey* in 1915 to become a military ferry at Chatham. It was said that amongst other duties, she was responsible for carrying the dispatches to Admiral Sir Roger Keys aboard a ship laying off the Nore in readiness for the raid on Zeebrugge on 23 April 1918.

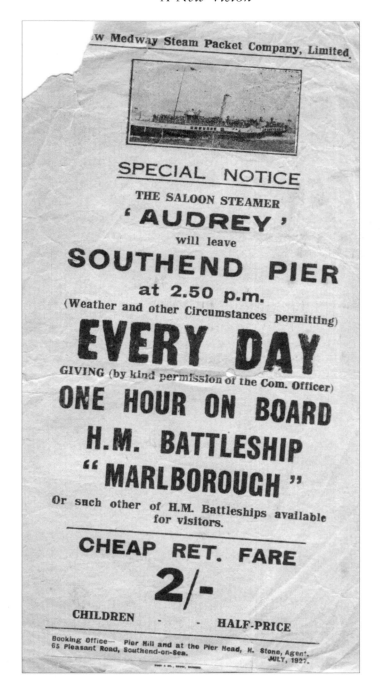

Handbill for cruises by *Audrey* from Southend to visit HMS *Marlborough* during July 1927. This was towards the end of *Audrey*'s career and she must have seemed very primitive to steamer enthusiasts, who were by then used to London River greyhounds such as General Steam Navigation's splendid new and distinctive *Crested Eagle*.

Audrey approaching Southend Pier around 1923. *Audrey*'s Herne Bay service in 1923 saw an early departure from Chatham's Sun Pier at 8.00 a.m. She then departed for Southend and left there at 10.00 a.m. for Sheerness to pick up and exchange further passengers. *Audrey* finally reached Herne Bay at 12.45 p.m. A return trip to Southend was then made with just three hours allowed before *Audrey* departed from Herne Bay for home. But, there was an added snag, as few boarded *Audrey* for the afternoon trip but huge numbers might join at Southend for the non-landing cruise. *Audrey* might therefore have around twice the ship's capacity to be transported back home in the evening! During 1923 and 1924, little *Audrey* worked a twelve-hour day for six days a week during the excursion season.

Herne Bay Pier was important to the New Medway Company during its 1920s period of expansion. A tramway was built on the pier in 1924. It was last used in 1939 and was sold for scrap after the war. The government gave £21,924 to the local council as compensation for wartime damage at the end of the 1939-45 war. *Medway Queen* was the last regular steamer to call at the pier, but *Queen of the South* made the final calls a few years later.

SOUTHERN RAILWAY
— AND —
New Medway Steam Packet Co., Ltd.

Daily Sea Trips

From Margate Jetty
(Fridays excepted) Weather and other circumstances permitting

To HERNE BAY

By Saloon Steamer "AUDREY"

Leave Jetty 10.30 a.m. — Arrive Back 12.45 p.m.
„ „ 2.45 p.m. — „ „ 5 p.m.
(Thence to Ramsgate)

Cheap Return Fare 2/6   Children 1/6
(Not Landing at Herne Bay)

Also Combined Boat and Rail Bookings as follows.
Times of leaving as above.

		Single		Return
FARES :—To Herne Bay	...	2/3	...	3/-
„ Children	...	2/-	...	2/4
Return by Afternoon Boat or Train same day.				
„ To Ramsgate (5 p.m.)...		2/-	...	2/3
„ Children	...	1/9	...	1/10
Return by Train only.				

LUNCHEONS, TEAS, REFRESHMENTS served on Board.

Further Particulars at Company's Booking Office at Extension
end of Jetty.   Local Agents : ACOCK & SON, Ramsgate.

Clarke & Knapp, Bilton Square, Margate.

Handbill for cruises by *Audrey* from Margate Jetty to Herne Bay during the 1925
summer season. Cruises were run in conjunction between the New Medway Steam Packet
Company and the Southern Railway as combined steamer and train fares were offered.
Audrey took up this service due to the appearance of the new *Medway Queen* and *Queen
of the South* in the previous year. The service between Margate, Herne Bay and Ramsgate
was a new venture for the company.

The small *Audrey* could accommodate 600 passengers and had a modest speed of 12 knots. Herne Bay had always been regarded as the Cinderella of the Thames resorts. Belle Steamers had though made occasional visits. Shippick saw the potential in opening up new calling points and Herne Bay was ideal for this purpose. Shippick placed Captain Tommy Aldis as master of the little *Audrey* on the Herne Bay route.

Southend Pier was a magnet for those interested in the New Medway Company steamers during the 1920s and 1930s. *Queen of Kent*, *Queen of Thanet*, *Medway Queen*, *City of Rochester* and *Essex Queen* would make often regular calls on their short trips to places like Herne Bay or on cross-Channel services.

Handbill advertising cruises by the *City of Rochester* between Southend, Herne Bay and Chatham during the 1925 season. Most services during the 1924 season operated from Sun Pier at Chatham. Steamers also lay off there at night.

Southern Railway and the
New Medway Steam Packet Company, Ltd.

THE SALOON STEAMER
"CITY OF ROCHESTER"
WILL RUN DAILY
(FRIDAYS EXCEPTED)
(weather and other circumstances permitting) between

HERNE BAY, SOUTHEND & CHATHAM.

Arriving from Southend	11.30 a.m.
Leaving Herne Bay	11.45 a.m.
Arriving at Southend	1. 0 p.m.
Returning from Southend	3. 0 p.m.
Arriving at Herne Bay	4.30 p.m.
Leaving Herne Bay	4.45 p.m.
Arriving at Southend	6.15 p.m.
Arriving at Chatham	8.15 p.m.

FARES.

Herne Bay to Southend **3/-** Return ; Single **2/3**. Child **1/9** Single or Return

Long Circular Tour, Herne Bay, Southend and Chatham (Boat and Rail) **4/6** Child **2/9** Return.

Circular Tour, Herne Bay and Chatham. **4/-** Return : Child **2/-** (Boat & Rail)

These Fares do not include Free Re-Admission to the Piers, for which 1d. each Passenger is charged at Sheerness, and 2d. at Southend and Herne Bay.

"Queen of the South" leaves Southend 5 p.m., arrives Chatham 7 p.m.

Train leaves Chatham for Herne Bay, Whitstable, Faversham and Margate

No Dogs allowed on Board. The Company do not guarantee exact time, but will do their best to ensure punctuality.

Luncheons, Teas and other Refreshments supplied on Board at Moderate Prices.

Office—280 High Street, Rochester.   Telephone—Chatham 353.

PERCY W. CAREY,
The Pier, Herne Bay.
Phone 22.

S. J. SHIPPICK,
Managing Director.

STANBROOK & SONS, "PRESS" OFFICE, HERNE BAY.

Audrey was sold to T. Ward & Company of Grays for £500 for scrapping in 1929. It was a sad end for a special pleasure steamer that helped to transform River Medway steamer services.

Medway Queen was one of the most fondly loved of River Medway paddle steamers. She was built by Ailsa of Troon in 1924. She was the first 'Queen' to be built for the newly advertised Queen Line of Steamers. She is shown here towards the end of her career.

Medway Steam Packet Co., Ltd.

TRAVEL BY THE
MEDWAY QUEEN
TO

FELIXSTOWE & WALTON

1½ Hours Ashore   3 Hours Ashore

2/-   Depart   1/3
CLACTON
11.55 a.m.
Arrive back
4.15 p.m.

RETURN.   CHILDREN HALF-FARE.   RETURN.

EVERY
MON., TUES., WED., THURS., FRID. (during AUG.) & SAT.

(Weather and other circumstances permitting)

WHY   You should travel by the Medway Queen.

FACTS !

1. It is the newest Steamer running on the East Coast, built 1924. Speed 18 knots and carries over 1,000 Passengers.
2. The only Boat that takes you to the L.N.E.R. Co. Pier, Felixstowe, close to BRITAIN'S LARGEST SEA-PLANE BASE where flights can be seen daily of the latest machines and practice flights for the coming Schneider Cup Race.
3. NO 2d. Pier Tolls are charged to our Passengers at the L.N.E.R. Pier, Felixstowe for re-admission to Pier. WE PAY THIS.
4. The L.N.E.R. Pier, Felixstowe, is only about 50 yards long, not half-a-mile like the new Pier, therefore no 2d. Tramfare to pay to reach the Shore End or no mile walk.
5. We can ALWAYS land you comfortably at the L.N.E.R. Pier, Felixstowe, no matter how rough the weather.
6. Buses meet the Boat at the L.N.E.R. Pier, Felixstowe and convey Passengers to any part of Felixstowe in 6 minutes.
7. A JAZZ BAND plays on BOARD during the Trip.

THE ONLY STEAMER ALLOWING 1½ HOURS ASHORE AT FELIXSTOWE.

DO NOT BE MISLED BY CATCH ADVERTS.
Ask for the Medway Queen that GIVES you FAIR SERVICE.

The Catering is in the hands of the Co., and excellent meals can be obtained on board, and Wines, Spirits and Beers at all times.

Clacton Agent : E. H. JONES, 292 Old Rd., Clacton.   Head Office : Rochester.
August, 1928.   Telephone : Chatham 2053.   S. J. SHIPPICK, Managing Director.

An early *Medway Queen* handbill issued in August 1928 shouting loudly to potential passengers the advantages of travelling by the newest paddle steamer in the New Medway Steam Company fleet. This handbill shows that Sydney Shippick, who was by then Managing Director, was firmly in control of his own publicity machine. His boldness in publicising the benefits of his pleasure steamers must have sent shock waves towards his competitors. It also alerts you to the fact that competition was very strong at Clacton, Walton and Felixstowe at the time. The handbill warns the public not to be misled by other adverts and to travel aboard *Medway Queen* to receive fair service.

Medway Queen made her maiden voyage on 18 July 1924. She sailed from Chatham to Southend and Herne Bay with many VIPs aboard. On the following day, she took over the Strood to Southend twice-daily service.

New Medway Steam Packet Co., Ltd.

CHEAP TRIP TO MARGATE

and Back, by the

FAST SALOON STEAMER

Medway Queen

MEDWAY QUEEN IS THE FIRST BOAT
Direct to MARGATE

MEDWAY QUEEN Leaves SOUTHEND PIER
Daily (Friday Excepted)
At 9.50 a.m.

MEDWAY QUEEN ALLOWS ABOUT **5** HOURS ASHORE

MEDWAY QUEEN Cheap Return Fare
3/6 available for Season

MEDWAY QUEEN Saturday and Sunday
4/6 Return

MEDWAY QUEEN Will Run to MARGATE
on Fridays during August

MEDWAY QUEEN HAS A SPEED OF ABOUT
18 Knots per Hour

MEDWAY QUEEN Has REFRESHMENTS on Board
at MODERATE CHARGES

**Booking Offices : First Box Down Pier Hill Steps
and at the First Box on the Left at Pier Head**
H. STONE, Agent, 65 PLEASANT ROAD, SOUTHEND.

Reliance Printing Co. (1912) Ltd., Tylers Avenue, Southend-on-Sea,

An early handbill for *Medway Queen* displaying aggressive advertising of cruises from Southend to Margate. Instead of placidly describing the attractions of *Medway Queen*, the New Medway Steam Packet Company proudly states the features that made *Medway Queen* stand out from her opposition. This is typical of the brave and inspired new company that Sydney Shippick had created on the River Medway.

The master of *Medway Queen* on her maiden voyage was Captain C. Scott. He was succeeded by Captain Aldis for the first two seasons. *Medway Queen* undertook the Strood to Southend service during her first season. In 1925, she took up the new service between the Medway towns and Southend to Clacton, Walton and Felixstowe. This route had been partially opened by *City of Rochester* in 1924 but only as far as Clacton.

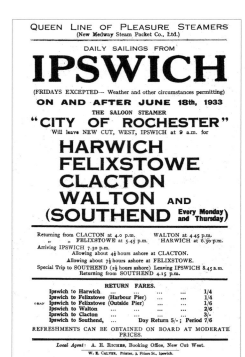

QUEEN LINE OF PLEASURE STEAMERS
(New Medway Steam Packet Co., Ltd.)

DAILY SAILINGS FROM

IPSWICH

(FRIDAYS EXCEPTED— Weather and other circumstances permitting)

ON AND AFTER JUNE 18th, 1933

THE SALOON STEAMER

"CITY OF ROCHESTER"

Will leave NEW CUT, WEST, IPSWICH at 9 a.m. for

HARWICH
FELIXSTOWE
CLACTON
WALTON AND
(SOUTHEND Every Monday and Thursday)

Returning from CLACTON at 4.0 p.m.   WALTON at 4.45 p.m.
        „   „   FELIXSTOWE at 5.45 p.m.   HARWICH at 6.30 p.m.
Arriving IPSWICH 7.30 p.m.
            Allowing about 4½ hours ashore at CLACTON.
            Allowing about 7½ hours ashore at FELIXSTOWE.
Special Trip to SOUTHEND (2½ hours ashore) Leaving IPSWICH 8.45 a.m.
            Returning from SOUTHEND 4.15 p.m.

RETURN FARES.

Ipswich to Harwich	...	...	...	1/4
Ipswich to Felixstowe (Harbour Pier)	...	...	1/4
Ipswich to Felixstowe (Outside Pier)	...	...	1/6
Ipswich to Walton	...	...	...	2/6
Ipswich to Clacton	...	...	...	3/-
Ipswich to Southend,	...	Day Return 5/- ; Period 7/6

REFRESHMENTS CAN BE OBTAINED ON BOARD AT MODERATE
PRICES.

Local Agent : A. E. RICHES, Booking Office, New Cut West.

W. E. CALVER, Printer, 3, Friars St., Ipswich.

City of Rochester handbill offering cruises from Ipswich in 1933. The last time that *City of Rochester* went down the River Orwell was on Sunday 27 August 1939 under the command of Captain Leonard Horsham. The steamer was extremely busy, as people wanted to have a cruise, as war was imminent. *City of Rochester* was recalled to her home in the Medway on the following day. It was a historic day, as it was the last time that a paddle steamer would grace the Orwell until *Waverley* during the 1990s.

1927 saw the New Medway Company purchase two ex-minesweepers named *Melton* and *Atherstone*. They renamed them *Queen of Thanet* and *Queen of Kent*. *Queen of Kent* made her maiden voyage trip on Sunday 29 July 1928. She departed Chatham at 7.30 a.m. and called at Southend and Margate before crossing to Calais. She made the same journey every day during that season with an off-service day of Saturday instead of the more usual Friday. This was because holidaymakers at the resorts were occupied in arriving and departing on their holidays and therefore didn't want to cross the English Channel to Calais.

Queen of Kent or *Queen of Thanet* (right) and *Golden Eagle* (left) alongside Margate Jetty around 1930. Margate Jetty during the late 1920s provided great entertainment in the middle of the day with several paddle steamers arriving and departing in quick succession. The jetty had three berths. *Medway Queen* started running to Margate in 1926. She was usually the first to arrive and occupied the best berth on the side of the pier head.

1925 allowed a more regular service for the steamers of the New Medway Company fleet now that there were two new steamers. *Medway Queen* was placed on the Clacton run and then onwards to Walton and Felixstowe. She departed at 8.30 a.m. from Chatham and arrived at Felixstowe at 1.15 p.m. After 1 hour 45 minutes ashore, she returned to Chatham, arriving back at 7.45 p.m. *Medway Queen* in shown here later in her career.

Queen of Thanet arriving at a Kent resort around 1935. *Queen of Kent* and *Queen of Thanet* both cruised to Margate. Once there, one would proceed to Calais and the other would cruise to Dover or would terminate at Margate and would then offer a sea trip. Several trial trips were also made that season to Boulogne, as the Medway Company had their eyes on it as a future destination.

Queen of Thanet at Ramsgate. *Queen of Thanet* made her maiden cruise to Calais on Sunday 19 May 1929. A few years later, those passengers wishing to travel from the Medway towns to Boulogne and Calais would embark on a special 'Sunshine Coach' to join the steamer at Margate.

Passengers aboard the *Queen of Thanet* in June 1935. Photographs such as this give us a real feel for the atmosphere of the time. *Queen of Thanet* could carry up to 871 passengers. Entertainment was often a popular feature of cruises. Passengers were also able to pose for a photograph with the Queen Line lucky black cat. Did you ever have your photograph taken with the cat?

Queen of Thanet moored on the Medway close to Chatham town centre. The New Medway Company wanted to expand their services on the long-haul routes, especially to the Continent. *Queen of Thanet* and *Queen of Kent* provided an initial answer to their needs. The expense to refit the two ex-Royal Navy minesweepers would be significantly less than building new tonnage.

One of the two-funnelled New Medway steamers at Margate Jetty along with another paddle steamer around 1935. Captain Aldis became master of *Essex Queen* in 1926 before moving to *Queen of Kent* in 1928, when the inaugural service to Calais was started. Aldis then moved to *Queen of Thanet* in 1929 and remained in command for six summer seasons. Tommy Aldis was joined by the young Leonard Horsham as First Officer. The happy and pleasant personality of Tommy helped to give *Queen of Thanet* the name of 'The Happy Ship'.

An advertisement for Queen Line Steamers around the early 1930s. In 1931, *Queen of Kent* and *Queen of Thanet* were converted to burn oil instead of coal and were fitted with a new design of paddle wheel that vastly increased the speed of each steamer. They weren't perhaps the most attractive paddle steamers in the fleet, but did serve the company well for almost twenty years.

140 MILES for 4/- !
GLORIOUS SEA CRUISES to CLACTON
Also to SOUTHEND, HERNE BAY, MARGATE, etc..

by QUEEN LINE STEAMERS

from LONDON BRIDGE WHARF, GREENWICH & N. WOOLWICH.

Full Particulars from Queen Line Agents at above piers.

Advertising by Stilwell, Darby & Co.Ltd., Charing Cross House, Charing Cross Rd., London, W.C.2.

A view aboard *Queen of Thanet* in 1935. *Queen of Thanet* was 235 feet long. She and her sister were very distinctive paddle steamers due to the distance apart of their funnels and their cruiser sterns. The sometimes huge number of passengers needing to return back from Herne Bay in the evening meant that a relief boat was sometimes needed. Sometimes, *Queen of Thanet* or *Queen of Kent* were diverted to call at Herne Bay to pick up excess passengers after their Calais trips. On occasions, they sometimes cut corners off the Kentish flats to make up time. Frequently, this led to passengers having to spend all night marooned on a sandbank, leading to the Calais service being renamed the 'no passport day and night trip'!

THE "ROYAL EAGLE" AND THE "QUEEN OF THANET" AT MARGATE.   9064.

Queen of Thanet (right) and *Royal Eagle* (left) alongside Margate Jetty around 1935. Medway favourite Captain Tommy Aldis became berthing master at Margate Jetty in 1937 due to his unhappiness with the amalgamation with General Steam Navigation. But as he said at the time, a 'landlubber's job was not for him'. He then took over command of the new *Royal Sovereign*.

The condition of *Queen of Kent* had deteriorated from her cross-Channel work and by 1934 was suffering from this daily routine. She was still capable of her role, but unreliability and passenger dissatisfaction was becoming an issue. This prompted the New Medway Company to place the order for *Queen of the Channel* – an altogether different breed of ship. This reduced the role of *Queen of Kent* to relief steamer and for services such as this from Ramsgate and Margate to Southend during 1935. With the debut of *Royal Sovereign* in 1937, *Queen of Kent* was quickly withdrawn.

Passengers aboard *Queen of Thanet* in June 1935. One of the most important developments of the New Medway Company was the advent of 'No Passport' trips to the Continent. In the days before the First World War, passports weren't required for cross-Channel trips. You simply turned up, bought a ticket and went on the cruise!

A scene aboard *Queen of Thanet* in 1928. Passengers are watching crew members working on the winch machinery.

QUEEN LINE of PLEASURE STEAMERS

New Medway Steam Packet Co., Ltd.

ON AND AFTER WHITSUN HOLIDAYS
CHEAP SEA & RIVER TRIPS

BY THE FAST SALOON STEAMERS.

"QUEEN OF THANET", etc., etc.

(weather and other circumstances permitting)

DAY	DESTINATION	STROOD	LEAVING Sun Pier	Gillingham	RETURNING
~~May 16th & 17th~~ May 15th & 16th	**MARGATE & CALAIS**		7.45 a.m.		Calais 4.15 p.m.
May 14th, 15th & 16th	**HERNE BAY MARGATE**		8.30 a.m.	8.45 a.m.	Margate 4.0 p.m. Herne Bay 5.0 p.m.
May 14th, 15th & 16th	**SHEERNESS SOUTHEND CLACTON FELIXSTOWE**	9.10 a.m. 3.15 p.m.	9.30 a.m. 3.30 p.m.	9.45 a.m. 3.45 p.m. (Sheerness & Southend Only)	Felixstowe 3.30 p.m. Clacton 4.30 p.m. Southend 11.15 a.m. and 5.0 p.m. Sheerness 11.45 a.m. and 5.30 p.m.
May 17th, to June 4th	**SHEERNESS SOUTHEND HERNE BAY MARGATE CLACTON FELIXSTOWE**		8.45 a.m.	9.0 a.m.	Felixstowe 3.30 p.m. Clacton 4.30 p.m. Margate 4.0 p.m. Herne Bay 5.0 p.m. Southend 6.30 p.m. Sheerness 7.0 p.m.

The Full Summer Service will come into operation on and after 4th June.

On May 18th and Every Wednesday following Steamer leaves Chatham for London (Tower Pier) at 9.0 a.m. Return Fare 4/6.

RETURN FARES—

Sheerness 1/6 (Boat only), 2/3 (Boat and Rail), Southend 3/-, Clacton 5/-, Felixstowe 7/-, Herne Bay (Boat only) 3/-, (Boat and Rail) 4/- Margate 3/6, Calais 10/6

(Children at reduced prices).

IMPORTANT NOTICE.
  Tickets are available for Season, excepting Calais, where Return Tickets are available on day of issue as shown. Passengers to France must be able to satisfy the Purser that they are British or French subjects, or they will not be permitted to land there. Passengers having gone ashore are not allowed on board again until half-an-hour before the advertised time of sailing. Continental Fares include French Landing Tax. Calais Casino— Admission Free to passengers on production of Steamer Voucher, issued on board. Parties catered for at reduced rates. Breakfasts, Luncheons, Teas, and Light Refreshments of the best quality can be obtained on board at moderate Prices.

MAY, 1932.   HEAD OFFICE, ROCHESTER.   TELEPHONE : CHATHAM 2053.   S. J. SHIPPICK, Managing Director,

Wood & Co., Printers, Brook, Chatham.

Handbill for Queen Line Pleasure Steamers from May 1932. With the acquisition of the *Queen of Kent* and *Queen of Thanet*, the New Medway Steam Packet Company decided to resurrect day excursions to France, which hadn't been run since the GSNC had sold *Kingfisher* at the end of the 1911 season. These day excursions quickly became very popular and offered a cheap fare of 8s 6d return between Margate and Calais.

A Queen Line steamer moored next to another ship, silhouetted at sunset and viewed from Chatham Sun Pier during the mid-1930s.

Queen of Kent and *Queen of Thanet* looked almost identical. This postcard titled *Queen of Kent* is identical to one titled *Queen of Thanet*! Two identical cards were printed. One had *Queen of Kent* written on the bows and the other had *Queen of Thanet*.

Queen of Kent emerged after her rebuilding on the Medway in 1928. Her role was exclusively that of a cross-Channel steamer. Unfortunately, *Queen of Kent* often only achieved meagre speeds whilst crossing the channel. This might be as low as 10 knots. Unfortunately, this ensured that time ashore on the Continent could be as short as fifteen to twenty minutes!

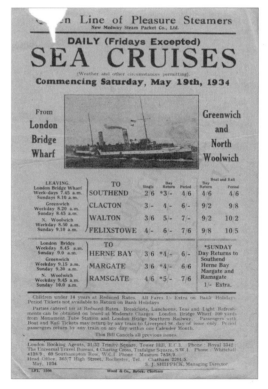

Handbill for cruises by Queen Line from London Bridge Wharf commencing May 1934. 1933 saw the New Medway Steam Packet Company operate two daily services from London. One was to Southend, Herne Bay, Margate and Ramsgate, and the other went to Southend, Clacton, Walton and Felixstowe. The Queen Line steamers departed from Fresh Wharf close to London Bridge. The usual New Medway steamers to be used on these routes were *Queen of Southend* and *Essex Queen*. Both were once owned by Belle Steamers, who operated from the same nostalgic location in the days before the First World War. Queen Line Steamers later operated from Tower Pier from around 1935 onwards.

An aerial view of *Queen of Thanet* at sea. Captain Aldis was born in 1897 and commanded *Queen of Thanet* at the end of the Second World War, later taking command of the newly built second motor ship *Queen of the Channel* in 1949. He retired some years later.

THE POPULAR "QUEEN OF KENT."

P.S. "QUEEN OF KENT."

One of the Queen Line Fleet of Oil Burning Pleasure Steamers operating on the Southend—Margate route.

Length 235 feet.   *Speed 16 knots.*   *Gross tonnage 798.*

QUEEN LINE STEAMERS—SERVICES INCLUDE :

London, North Woolwich, Greenwich, Tilbury, Gravesend, Chatham, Rochester, Strood, Gillingham, Sheerness, Southend, Clacton, Walton, Southwold, Lowestoft, Yarmouth, Felixstowe, Harwich, Ipswich, Herne Bay, Margate, Ramsgate, Broadstairs (via Margate), Calais, Boulogne, Ostend, etc.

Advertisement from a Queen Line brochure of the mid-1930s. It's impressive to read the destinations covered by the fleet at the time.

Men painting the paddle box of *Queen of Kent* on the Medway. *Queen of Kent* was a well-appointed steamer and could comfortably transport her enthusiastic passengers. Very often, she would record full sailings for several days in a row.

Work being carried out on the exposed paddle wheel of *Queen of Kent* around the time of her acquisition by the New Medway Company in 1928. At the time of her acquisition for Medway service, she was virtually rebuilt internally.

Queen of Southend approaching Southend Pier around 1931 prior to her refit. At the time, her route was usually Chatham to Felixstowe. During her 1931 refit, she was converted to burn oil.

THE PLEASURE STEAMER

'Queen of Southend'

(WILL RUN DAILY, FRIDAYS EXCEPTED.)

Margate, Ramsgate, Southend and London.

	Weekdays	Sundays
Leaving **London Bridge Wharf**	8.45 a.m.	9. 0 a.m.
Leaving **Southend** ...	11.45 a.m.	12 noon.
Leaving **Herne Bay** ...	1. 0 p.m.	1.15 p.m.
Arrive **Margate** ...	2. 0 p.m.	2. 0 p.m.
Arrive **Ramsgate** ...	2.45 p.m.	2.45 p.m.
Depart **Ramsgate** ...	2.45 p.m.	2.45 p.m.
Depart **Margate** ...	3.45 p.m.	3.45 p.m.
Depart **Herne Bay** ...	4.45 p.m.	5. 0 p.m.
Southend, Change for Chatham	6. 0 p.m.	6. 0 p.m.
Gravesend	7. 0 p.m.	7. 0 p.m.
North Woolwich ...	8. 0 p.m.	8. 0 p.m.
Greenwich	8.30 p.m.	8.30 p.m.
London Bridge Wharf Arrive ...	9. 0 p.m.	9. 0 p.m.

Fares :- Ramsgate Day Return 3/- Period 5/- Boat and Rail 3/6 Child 1/6 Single or Return. Gravesend ; Single 3/3 Period 5/6 London ; Single 3/6 Period 6/- Child ; Single 2/- Period Return 3/-

TO MARGATE & SOUTHEND FRIDAYS

Leaving **Southend** ... (about)	...	10.40 a.m.
Arriving from **Southend** ... ,,	...	11.55 a.m.
Leaving **Herne Bay** for **Margate** ,,	...	12. noon
Arriving at **Margate** ... ,,	...	1. 0 p.m.
Returning from **Margate** ... ,,	...	4.30 p.m.
Arriving at **Herne Bay** ... ,,	...	5.25 p.m.
Leaving **Herne Bay** for **Southend** ,,	...	5.30 p.m.
Arriving at **Southend** ... ,,	...	6.35 p.m.

Herne Bay to Margate 2/- Return. Boat & Rail 2/6 Single 1/6 Child 1/3 Single or Return. Boat & Rail 1/6

The Company do not guarantee exact time, but will do their best to ensure punctuality.

PERCY W. CAREY, (Agent,)   S. J. SHIPPICK,
35 Bank Street, Herne Bay, Phone : 318.   Managing Director.

18/8/33.   The Telford Printing Works, 10 Bank Street, Herne Bay.

Cruises by *Queen of Southend* are advertised on this 1933 handbill. Her service from London to the Kent coastal resorts was quite punishing, as can be seen from this handbill.

Southend Pier during its heyday. During the early 1920s, Southend Pier was a wonderful place to be on a Saturdays between 11.30 and 12.30 and 5.30 to 6.30 when paddle steamers could be seen arriving and departing from the River Medway and London. Before the Prince George Extension was opened in 1929, there was only one berth and up to seven paddle steamers could arrive in quick succession. It was usual for two or three to queue, and *City of Rochester* from Strood and Chatham often sneaked in between the others.

THAMES QUEEN

Queen of Southend had her promenade deck extended up to the bow in 1936 at the New Medway Company's yard at Rochester. This altered her appearance dramatically from her earlier look.

Queen of Southend could carry up to 976 passengers. She was renamed *Thames Queen* in 1938 as a consequence of the merger between the New Medway Company and General Steam Navigation.

Queen of Southend approaching Felixstowe during the mid-1930s. She lasted until early 1947, when she was sold for scrap to Metal Industries Limited.

Rochester Queen was purchased by the New Medway Company after the 1931 season. She was a very modest vessel, once having been the Gravesend to Tilbury ferry named *Gertrude*. She was placed on the twice-daily service between Rochester and Southend.

Royal Daffodil in the River Mersey during the First World War. At the time of her acquisition, she was the smallest steamer in the fleet. She had a fine pedigree, having taken part in the famous Zebrugge raid in 1918. At the time, she had been a River Mersey ferry named *Daffodil*, but George V commanded that, for the rest of her career, she be known as *Royal Daffodil* in recognition of her deeds. A metal plaque could be seen explaining her gallantry aboard the steamer. Her funnel and upperworks also showed the marks that shrapnel had made during the raid.

Pleasure  Steamers
(New Medway Steam Packet Co., Ltd.)

Come and See Your Navy.

SATURDAY, AUGUST 4th to SATURDAY, AUGUST 11th.
(Sunday Excepted)

THE FAMOUS TWIN-SCREW STEAMER
"ROYAL DAFFODIL"

This Steamer is the famous Daffodil which made history in the raid on Zeebrugge in 1918, and to which His Majesty The King graciously gave a plaque stating that the name should be in future "Royal Daffodil."

Will leave Southend Pier at 2.30 p.m. for

H.M.S. VALIANT
OR
H.M.S. RENOWN

Giving 2 hours aboard
Again at 4.15 p.m. giving 1 hour aboard.

DON'T MISS THIS OPPORTUNITY.

Fare 1/6 Children 1/-

Luncheons, Teas and Refreshments (Fully Licensed) of the best quality can be obtained on board at moderate prices.

S. J. Shippick, Managing Director.   Local Agent : A. Oswald, The Pier, Southend.
Head Office : Rochester. Tel. 2204-5 Chatham.   Phone : Marine 67736.

S. SPARKES, 193 HIGH STREET, SOUTHEND-ON-SEA (ENTRANCE IN QUEENS ROAD)

Cruises by *Royal Daffodil* in 1934 to view *Renown* and *Valiant*. Such trips were commonplace at the time, and time aboard the royal naval ships was allowed.

The first *Royal Daffodil* on the River Medway around 1934. *Royal Daffodil* had a substantial promenade deck, as can be seen in this view. By this time, she was twenty-eight years old, having been built in 1906. *Royal Daffodil* underwent work on her bridge during the winter of 1933/34. During the 1934 summer season, she operated on the Medway towns to Southend service.

Royal Daffodil cruising off Southend during the mid-1930s.

Duchess of Kent being refitted on the River Medway as *Clacton Queen* around 1934. She had been purchased from the Southern Railway after service on the south coast. She had originally been built by Day, Summers & Co., of Southampton in 1897.

Duchess of Kent being refitted on the Medway for her new role of *Clacton Queen* around 1934. Once refitted, her time on the Medway was short-lived, as in November 1935, she was sold for further service at Blackpool.

Clacton Queen at Ipswich around 1935. She was a competent sea boat, despite being built for the calm waters between Portsmouth and Ryde. She often had to cope with some challenging seas on trips from Ipswich or Clacton to Chatham.

The New Medway Company's own newspaper *The Holiday Mail* from 1935. It offers a great snapshot into the thriving company at the time as well as the wide number of places visited. It also tantalises the reader with a sketch of the yet unbuilt *Queen of the Channel*.

ALL QUEEN STEAMERS ARE HAPPY SHIPS.

Three young women pose aboard the *Queen of Kent* in 1934. Competitions were held each season for passengers to take photographs like this of their trip. The winning photos were then included in the following year's edition of the newspaper. At the outbreak of the Second World War, *Queen of Kent* became a minesweeper once again.

Passengers disembarking from *Queen of Thanet* in 1935. It must have caused significant confusion when an inebriated passenger, after having a drink or two ashore, returned to board a paddle steamer to their embarkation point and was faced with a plethora of 'Queen' steamers!

Handbill advertising cruises to Cockleshell Beach aboard *Rochester City Belle* during the 1934 season from Chatham and Gillingham. 'Cockleshell' or 'Thames Beach' was situated on the Isle of Grain in the Medway and provided simple seaside pleasures for trippers from the Medway towns who were unable to afford to travel to large resorts such as Margate or Southend.

Handbill advertising cruises from the Admiralty Pier, Gillingham, and Chatham Sun Pier during the 1935 season. The upriver cruises passed through the picturesque Aylesford Bridge to Allington Lock with time ashore. Cruises were offered aboard the New Medway Company's motor launch *Rochester City Belle*. Other small river craft operated by the company included *HRH Princess Mary* and *The New Medway*.

Departing from Southend with a full load of passengers during the 1930s. The head office of the Queen Line of Pleasure Steamers was located at 365 High Street in Rochester.

City of Rochester departing from Southend Pier around the late 1930s.

Cover for the Queen Line of Pleasure Steamers brochure dating from around 1934. It shows a striking colour image of *Queen of Thanet*, which was one of the main vessels owned by the company. Along with the usual history of the company, the brochure includes a full guide to the River Thames as well as listing a number of local Medway businesses who supplied the company with provisions. These range from Jasper & Sons, the bakers, Jenner's Ales and Phillips & Pett, the mineral water manufacturers.

22099 Chatham. S.S. "Princess of Wales"

Princess of Wales arriving at Chatham Sun Pier with St Mary's church, the Royal Marines Barracks and the Royal Naval Dockyard in the distance.

Chatham Dockyard, from Upnor

Chatham Dockyard viewed from the hills above Upnor around 1905. *Princess of Wales* is alongside Upnor Pier ready to depart on a cruise. Steamer passengers would often disembark and then walk up through the woods to have a picnic on Upnor Hill. There, they had a fine view of the dockyard beyond. At Upnor, there were also some attractions, such as tea gardens.

Queen Line of Pleasure Steamers brochure published in 1937. The New Medway Steam Packet Company produced high-quality and visually stunning material promoting its services from Rochester, Chatham and London, etc. This particular brochure heavily pushes the modernity of the new motor ships and the significant changes achieved by the company from a generation earlier. It includes information on the Medway steamer fleet, places visited by the vessels and how to get the most out of a Queen Line cruise. It included everything that you could need for a splendid Queen Line cruise for just 3d.

A May 1930 handbill for *Medway Queen* offering cruises from Southend to Herne Bay and Margate. *Medway Queen* was able to use the extensive new Prince George Extension opened during the previous season. By now, the New Medway Steam Packet Company were proudly calling themselves the Queen Line of Pleasure Steamers.

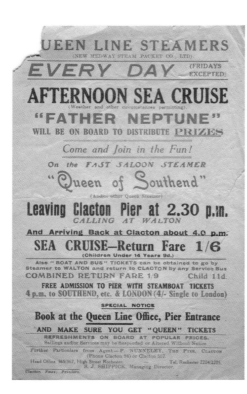

Handbill advertising sea cruises by *Queen of Southend* from Clacton Pier around 1930. The handbill advertises the fact that Father Neptune would be aboard to distribute prizes. The company employed several such novelties to encourage passengers to board their steamers rather than those of their competitors. By this time, tickets were also available offering a combined boat and bus option.

Medway Queen during her heyday cruising in the Thames Estuary. From 1937, the combined fleets of the New Medway Company and General Steam Navigation Company became officially known as Eagle & Queen Line steamers after GSN bought up all of the share capital of its Medway-based rivals.

Chatham Navy Week was always an extremely popular event. This wonderfully patriotic handbill dates from July 1934. The evening cruise took passengers from Sun Pier to Southend and Sheerness to view warships at anchor. Such cruises were obviously popular and joyous events and dancing was allowed on board the steamers.

Medway Queen departing from Southend Pier in 1963. During that season, she operated from 1 June until 8 September. *Medway Queen* day return fares during the 1963 season rose by 6d or 1s. Catering prices, though, were dropped due to a review by the operating company. Meals cost 3s 6d less and lunch was 2s cheaper.

QUEEN LINE PLEASURE STEAMERS

NEW MEDWAY STEAM PACKET COMPANY, LTD.

SPECIAL TRIPS
TO
HERNE BAY
AND
MARGATE

MONDAYS, TUESDAYS, WEDNESDAYS, THURSDAYS & SATURDAYS (SUNDAYS 4/6)

3'6

DAY RETURN - CHILDREN UNDER 14 HALF FARE

STEAMER LEAVES GREENWICH 9.15 a.m. (SUNDAYS 9.30 a.m.)
" " N.WOOLWICH 9.45 a.m. (SUNDAYS 10.0 a.m.)
Weather and other circumstances permitting

ABOUT 4 HOURS ASHORE IN HERNE BAY
" 2 " " " " MARGATE

Parties catered for at reduced rates. Breakfasts, Luncheons, Teas and Refreshments can be obtained on board at moderate charges
Sailings and/or Services may be may be altered or suspended without notice.
Head Office : 365 367 High Street, Rochester.   Phone : Chatham 2204/5.   S. J. SHIPPICK, Managing Director
2929   Wood & Co., Brook, Chatham.   AUG., 1935

Handbill advertising Queen Line cruises from Greenwich and North Woolwich to Herne Bay and Margate during August 1935. Fares aboard the New Medway Company's Queen Steamers were often cheaper than those aboard the GSN steamers. This might be as much as 30 per cent. But, sometimes, the older Medway fleet were less reliable with minimal time ashore or extremely late arrival back at the departure point.

An evocative view of the promenade deck of the *Medway Queen* during her final years in service. *Medway Queen* was built by Ailsa of Troon in 1924 and had paddle boxes and vents characteristic of Ailsa's work. She was built to the specification of Captain Shippick of the New Medway Steam Packet Company. Originally, she was coal-fired but was later re-boilered and converted to oil in the late 1930s.

Medway Queen.

An unusual artist-designed image of *Medway Queen* produced as a postcard towards the end of her career. She is shown having just departed from Strood Pier with the familiar skyline of Rochester castle, cathedral and bridge in the distance.

Preliminary.

THE QUEEN STEAMERS

(New Medway Steam Packet Company, Limited).

Early Season (1935) sailings from MARGATE PIER commencing 23rd May

From May 23rd to May 30th inclusive (but excluding Friday).	From June 1st onwards. Every day, Fridays included
A SPECIAL AFTERNOON TRIP TO SEA	**A SPECIAL AFTERNOON TRIP TO SEA**
Leaving Margate Pier at 2.20 p.m., and returning at about 4.10 p.m. Destination varied daily as advertised by special bills.	Leaving Margate Pier at 2.40 p.m., and returning at 4.40 p.m. (Fridays and Saturdays at 4.30 p.m.) Destination varied daily as advertised by special bills.
ADULT 1/9. CHILD (under 14) 10d. return.	ADULT 1/9. CHILD (under 14) 10d. return.

(SPECIAL TERMS TO PARTIES ON ALL THESE TRIPS).

On and after June 8th RAMSGATE. Leaving Margate Pier at 2.0 p.m. and arriving back at 3.45 p.m.

ADULT 1/9. CHILD (under 14), 1/- return.

NOTE.—Passengers may land at Ramsgate Harbour and return by any Granville Sunshine Saloon Coach, or by Tram, any time same day with the same ticket.

From May 23rd until June 6th, inclusive (but excepting Fridays)

Fast Service to SOUTHEND, GRAVESEND, NORTH WOOLWICH, GREENWICH & TOWER PIER

Leaving Margate Pier at 4.15 p.m. (Sundays 4.30 p.m.).

	Southend	Gravesend	Nth. Woolwich	Greenwich	Tower Pier
Adult Single	3/-	3/3	4/-	4/-	4/-
Adult Period	5/-	6/-	6/6	6/6	6/6

Children at reduced rates on all trips

NOTE—On and after June 8th (Whit. Saturday), this London service will leave Margate Pier at 3.45 p.m. and will terminate at Greenwich.

June 1st to June 6th (inclusive) HERNE BAY. Leaving Margate Pier Weekdays at 4.15 p.m., Sundays 4.30 p.m.

Adult Single 1/6. Children at Reduced Rates.

On and after June 8th, the Herne Bay Service will leave Margate at 3.45 p.m. (Fridays 4.30 p.m.)

SPECIAL TRIPS TO FRANCE

Commencing June 9th, until further notice, Every SUNDAY, TUESDAY and THURSDAY. Day trips to CALAIS, (An extra trip will be run on WHIT MONDAY). Leaving Margate Pier at 11.30 a.m., allowing about 2½ hours ashore in France and arriving back at Margate at about 6.45 p.m. (No Passports Required).

ADULT RETURN 8/9. CHILD (under 14) 5/-.

Fares include French landing tax and admission to Calais Casino.

No dogs, bicycles or luggage allowed on these trips to Calais. Passengers must be British, French or Belgian subjects, or they will not be permitted to land there. If at any time, for any reason whatsoever, the vessel does not journey to destination, passengers will be charged the amount which is proportionate to the distance covered.

On all Queen steamer journeys, other than to the Continent, passengers are allowed luggage up to 100lbs, per person, free.

All refreshments and first class catering on board at popular prices. All sailings, weather and other circumstances permitting.

WATCH FOR LATER NOTICE OF REGULAR SAILINGS TO OSTEND, BOULOGNE and CALAIS, by the new Wonder Ship "Queen of the Channel" and for the "Long day at the seaside" service, by the new Super Luxury Liner, "Queen of the Channel," commencing 29th June.

Issued by the New Medway Steam Packet Co., Ltd., Rochester.   S. J. Shippick, Managing Director.
W. S. Gabriel (Local Agent), Margate Pier and Ramsgate Harbour, Phone Margate 1430.

Wood & Co., Brook, Chatham.   May, 1935.

Handbill advertising cruises from Margate Pier during the Silver Jubilee celebrations of George V and Queen Mary in 1935. It notes that each passenger could carry 100 pounds in weight of luggage on a cruise.

Two young deckhands aboard *Medway Queen* watch the steamer tie up alongside Southend Pier around 1961. Astern of *Medway Queen* is the ill-fated *Anzio I*. She took over the Sheerness to Southend ferry service from her predecessor *Anzio* and could carry up to 351 passengers. Work at Sheerness meant that she could no longer call there after 1963, so she was laid up. Sadly, in 1966, she sunk off Spurn Head on the way to a new career.

Queen of the South manoeuvring in the River Medway close to Chatham's Sun Pier in 1966. *Queen of the South* was a beautiful but rare visitor to the River Medway. Her career on the Thames was short-lived due to mechanical problems and lack of experience. She was crewed by several New Medway Steam Packet veterans, including Captain Tommy Aldis and Captain George Fowle.

Queen of the South moored on the River Medway at Chatham viewed from the Riverside Gardens around 1965. *Queen of the South* had an unsuccessful time on the River Thames and was moored for a short while on the Medway. *Queen of the South* had arrived on the River Medway on her delivery voyage from the Clyde. It was stated that no other vessel overtook her. This was a testament to her build and condition.

Medway Queen on a barge off Chatham town centre soon after her arrival back on the River Medway during the early 1980s. Soon after, the *Medway Queen* Preservation Society was formed to restore her.

Medway Queen alongside the Victorian pumping station at Chatham Dockyard. The crane, offices and other buildings behind the pumping station were part of the nuclear refitting facility. Most have since been demolished. *Medway Queen* stayed at this location until towed to Damhead Creek in late 1987.

Kingswear Castle was restored on the River Medway at the Medway Bridge Marina at Borstal. In the few years after her withdrawal from service on the River Dart, the condition of *Kingswear Castle* deteriorated dramatically. Her restoration was lengthy and undertaken by a dedicated band of volunteers from the Paddle Steamer Preservation Society.

Kingswear Castle during her early days of service on the River Medway before she gained her full passenger certificate. Since those days, *Kingswear Castle* has changed greatly, reflecting the needs of the passenger and her new life on the Medway. These annual changes have made dating photographs easy!

Clyde was a familiar sight on the River Medway during the 1980s. She was built by the well-known builder A. & J. Inglis of Pointhouse in Glasgow in 1960 – the builders of the famous *Waverley*. She was built as a tug for the Clyde Navigation Trust and undertook occasional excursion work. After spells at Ullapool and Kyle, she eventually ended up being owned by Invicta Line Cruises on the River Medway in 1984. *Clyde* is shown here alongside Strood Pier.

Kingswear Castle in one of the dry docks at Chatham Historic Dockyard. Luckily for *Kingswear Castle*, the completion of the restoration project coincided with the opening of the Georgian and Victorian section of the former Royal Naval Dockyard as a leading national tourist attraction. This has enabled her to attract passengers who were visitors to the Historic Dockyard as well as showing them the impressive buildings and other Medway heritage only visible from the river.

When the summer season of cruises ends in October, *Kingswear Castle* then undergoes many months of work and surveys to prepare her for the following season. For several weeks each year, she is painted and repaired on the slipway at Strood. Here she makes a rare visit to one of the dry docks at the Historic Dockyard in April 1991. This dry dock is now occupied by *Cavalier*.

The River Medway has a new operational paddle steamer – the exciting moment when *Kingswear Castle* is in steam again in November 1983. She is seen passing close to the original Medway motorway bridge.

Chief Engineer Chris Smith prepares *Kingswear Castle* for another summer season on the River Medway by carefully painting her name on the bow.

Kingswear Castle's splendid paddle box. It shows not one of the famous Medway castles, but instead one from her native river – Kingswear Castle on the River Dart.

The beautifully restored *Kingswear Castle*. She now carries on the great tradition of the New Medway Steam Packet Company and its steamers.

PADDLE STEAMER KINGSWEAR CASTLE

* BAR   * EDWARDIAN PANELLED SALOONS   * AVAILABLE FOR PRIVATE HIRE
* BUFFET   * SOUVENIR SHOP   * 1986 STEAM HERITAGE AWARD WINNER

REGULAR CRUISES		From CHATHAM HISTORIC DOCKYARD	From STROOD PIER (Nr Rochester)	From SOUTHEND PIER
SUNDAYS 17 May - 6 Sept. (Except 31 May, 14, 28 June, 2 Aug.) **WEDNESDAYS** 22 July - 9 Sept. **FRIDAYS** 3 July - 11 Sept.	AFTERNOON MEDWAY CRUISES (2½ Hours)	2.30pm	3.00pm	—
THURSDAYS 9 July - 3 Sept.	SPECIAL 1½ HOUR THAMES CRUISES	—	—	11.00am, 1.45pm 3.15pm, 8.00pm
FRIDAYS 3, 17, 31 July, 14 August.	JAZZ JAMBOREES with the Premier Jazz Band	8.00pm	—	—
SATURDAYS 20 June, 11 July, 8,15,22 August & SUNDAY 11 Oct.	GREAT DAYS OUT (6 Hours)	11.00am	11.30am	—

FOR FULL DETAILS of this and our other steamers WAVERLEY and BALMORAL, 'PHONE (0634) 827648. Or write to: Kingswear Castle Excursions Ltd., Chatham Historic Dockyard, Chatham ME4 4TQ.
All tickets issued and passengers carried subject to the conditions of carriage of Kingswear Castle Excursions Ltd.

An early poster advertising cruises aboard the *Kingswear Castle*. Upnor Castle can be seen in the distance.

City of Rochester arriving at Felixstowe. Captain Horsham berthed the very last pleasure steamer to call at Felixstowe Town Pier on 27 August 1939, when he was master of the *City of Rochester*. At the outbreak of the Second World War, he commanded *Thames Queen*, which had been fitted out as an anti-aircraft ship.

Margate was a vitally important calling point for Queen Line Steamers. This 1936 handbill advertises an unusual cruise destination – Sandwich Bay. Passengers could view the North Foreland, Broadstairs, Ramsgate and Pegwell Bay during the cruise. The New Medway Steam Packet Company frequently employed gimmicks to encourage passengers to board their steamers. On this occasion, passengers were able to have their photograph taken on board and there was a free prize competition for all.

"QUEEN" STEAMERS

First Booking Office Margate Pier

TO-DAY
SPECIAL TWO HOURS
Long Sea Trip
By "QUEEN" STEAMER to
SANDWICH BAY
Passing Glorious Coastal Scenery
Leaving at **2.40** p.m. Returning at **4.40**

Adult **1/9** Return **10d** Child (under 14)
including Free Admission
to the Pier
Special Terms to Parties

Come with us on this delightful long Trip and see the Beauty of the Kent Coast, and the North Goodwin and Brake Lightships seawards.
The Trip also affords view of Kingsgate Castle, North Foreland, Broadstairs, Ramsgate, Pegwell Bay, Sandwich, and the Mystery Port of Richborough.
Your Photograph taken on board and Free Prize Competitions for all.

Be sure to take your Ticket at the "QUEEN" Steamer Office
W. S. GABRIEL (Local Agent).
New Medway Steam Packet Co., Ltd. 'Phone Margate 1430
COOPER, PRINTER, MARGATE.

By the end of the 1936 season, rivalry between the GSN and New Medway Company was becoming an issue. GSN was also aware of financial difficulties being experienced by the New Medway Company. A meeting was held at the GSN headquarters in London to discuss the issue of competition and fares. By the following season, a merger between the two companies was complete. Sydney Shippick remained as Managing Director of the now-semi-separate New Medway Company. The merger became a cost-cutting exercise that resulted in the reduction of agents around the Kent and Essex ports and a leaner timetable for 1937.

"Queen Of The Channel" Arriving, Margate Pier. 44.

The new *Queen of the Channel* arriving at Margate around 1935. Captain Tommy Aldis took the command of the splendid new vessel.

A portion of the palatial dining saloon aboard *Queen of the Channel*. Such facilities were bigger and better than other steamers. Look at the smaller dining saloon aboard *Medway Queen* to compare how revolutionary the *Queen of the Channel* was. *Royal Sovereign* later became a larger version of the ground-breaking *Queen of the Channel*.

Part of the vast Observation Lounge aboard Shippick's splendid new *Queen of the Channel*. When Shippick's dream came to fruition, passengers were able to enjoy better accommodation than could be obtained elsewhere on the Thames and Medway.

The revolutionary and first *Queen of the Channel* made her maiden voyage from the Southern Railway Pier at Gravesend on Saturday 22 June 1935 making calls at Southend and Margate before heading across the English Channel to Ostend. Passengers from the Medway towns were carried by Pilcher's 'Sunshine Coaches' to Margate to catch the ship for the Continent.

Queen of the Channel approaching Southend Pier on a sunny afternoon in 1938 as viewed from the Sun Deck of the famous pier. The pier head contained many fine attractions including the Dolphin Restaurant, cafés, bars, amusements and live music.

Queen of the Channel approaching the harbour at Ostend on 16 August 1937. Around a year earlier, the almost new Queen of the Channel was chartered to witness the maiden voyage of the Cunard liner Queen Mary from Southampton on 27 May 1936.

"QUEEN" STEAMERS

Booking Offices—Margate Pier and Ramsgate Harbour

Every Saturday and Monday
SPECIAL DAY TRIPS to

OSTEND

FINAL SAILINGS of SEASON To OSTEND
Saturday, 19th, and Monday, 21st September

ADULT		CHILD (under 14)
10/6	RETURN	**5/6**

THE SUPER-LUXURY MOTOR SHIP

"Queen of the Channel"

**Leaves Margate Pier for Ostend at 10.15 a.m.
on Saturdays and Mondays**

Allowing about 2 hours in Ostend and arriving back at Margate
at 6.45 p.m.

NO PASSPORTS REQUIRED
FIRST CLASS CATERING and all REFRESHMENTS AT POPULAR
PRICES.

DO NOT MISS THE OPPORTUNITY OF VISITING OSTEND BY THE
FINEST VESSEL OF HER CLASS AFLOAT
BOOK AT MARGATE PIER OR AT RAMSGATE HARBOUR

The New Medway Steam Packet Co., Ltd.   The Queen Line.
W. S. GABRIEL (Agent) Margate and Ramsgate.
Phones—Margate 1430 and Ramsgate 1080.

IMPORTANT NOTICE.—Passengers to Belgium must be British, French or
Belgian subjects or they will not be permitted to land there. Passengers having
gone ashore are not allowed on board again until one hour before the advertised
time of sailing. No luggage, dogs or bicycles allowed on day trips to Belgium. Fares
include Continental landing tax. If at any time, for any reason, the Company's
vessel does not complete the journey to destination, passengers will be charged an
amount proportionate to the distance covered.
Head Office : 365-67 High Street, Rochester.   Tel. : Chatham 2205-4.
                                            S. J. SHIPPICK, Managing Director.
(All sailings—weather and other circumstances permitting).

COOPER, PRINTER, MARGATE.

Handbill advertising cruises from Ramsgate and Margate to Ostend aboard the *Queen of the Channel* during the 1936 season.

Sydney Shippick's dream has come true. His splendid new Denny-built motor ships *Royal Sovereign* and *Queen of the Channel* at Ostend on 14 August 1937.

The first *Queen of the Channel* soon after entry into service. Shippick had created a particularly fine-looking vessel for the New Medway Steam Packet Company. She looked very much like the Clyde pleasure steamers, such as the *Queen Mary*. She was launched on 3 May 1935. She became the first diesel-engined cross-Channel steamer in Britain.

The new *Queen of the Channel*, pride of the Queen Line fleet, departing from Margate during 1937. Just a few short years after this photograph was taken, *Queen of the Channel* was bombed early in the morning of 28 May 1940. Her back was broken and all men aboard her were rescued. *Queen of the Channel*'s career was short-lived but her entry into service had been revolutionary.

The New Medway Steam Company's second ground-breaking motor ship was to be called *Continental Queen*, but before she came into service, the merger with General Steam Navigation had taken place. With GSN having the controlling interest, they dropped the proposed name and instead used one of their old favourites, *Royal Sovereign*.

Handbill for a special cruise departing from Strood Pier, Chatham Sun Pier and Gillingham Admiralty Pier to Southend with a cruise round the Nore lightship during the 1938 season.

BREAKFAST.

No. 1   **1/6**

Grape Fruit.
Tea or Coffee.
Bread and Butter.
Toast.   Preserves.

No. 2   **2/-**

Grape Fruit.
Tea or Coffee.
Bread and Butter.
Boiled or Poached Eggs.
Toast.   Preserves.

No. 3   **2/6**

Grape Fruit.
Tea or Coffee.
Bread and Butter.
Fish or Eggs and Bacon.
Toast.   Preserves.

No. 4   **3/6**

Grape Fruit.
Tea or Coffee.
Bread and Butter.
Fish.   Eggs and Bacon.
Toast.   Preserves.

EAGLE & QUEEN LINE STEAMERS.   **1938.**

Breakfast menu aboard Eagle & Queen Line Steamers during the 1938 season. The list of menu options was extensive, and this was matched by high standards of service and dining facilities – as good as anything you could find ashore in a fine hotel. Service was of course silver service and the company always used monogrammed silver-plated jugs, cutlery, coffee pots and teapots.

HOT LUNCHEON OR DINNER
(BY SPECIAL ARRANGEMENT).

No. 7	5/9

Soup, Thick or Clear.

Boiled Salmon.
Hollandaise Sauce.

Lamb Cutlets. Green Peas.
Roast Chicken. Watercress.
York Ham.

Various Vegetables
in Season.

Wine Jellies.
Vanilla and Strawberry Ices.

Biscuits.
Gorgonzola. Cheddar.
Coffee.

EAGLE & QUEEN LINE STEAMERS. 1938.

Menu for the special 'Number 7' hot luncheon or dinner offered by the Eagle & Queen Line during the 1938 summer season. The menu shows a liveried steward typical of those aboard Queen Line steamers at the time. The modernistic and sleek new motor ships with such a wealth of facilities would have amazed diners from a generation earlier aboard old and smaller paddle steamers.

Margaret Stone posing aboard *Royal Sovereign* in 1937. *Royal Sovereign* made her first trip to Southend Pier on Sunday 11 June 1937. Whilst alongside, she was opened to enable the public to inspect her. They weren't disappointed! Three days later, on Wednesday 14 June, she made her inaugural run from Gravesend to Ostend with calls at Southend and Margate. She was described as the first seagoing passenger vessel to have side 'blisters' to increase stability and space within her hull.

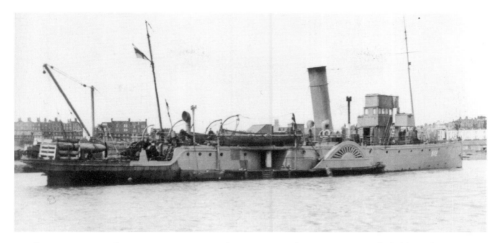

Medway Queen in her wartime guise early in 1940. The evacuation of children from London and North Kent utilised many of the peacetime pleasure steamers. This serious business resulted in some humour. For example, some children, after being out of sight of land during the trip, when asked where they came from when they landed at Lowestoft, answered 'England'. Another master who had left with a full complement of passengers found himself to be three over when disembarking. The answer was simple, three women had given birth during the trip!

HMS *Medway Queen*, departing from Harwich, passing through the ice floes during the winter of 1939/40. Several ships of the Medway fleet took part in the evacuation of Dunkirk. Whilst the exact number will probably never be known, it is thought that 1,650 were rescued by *Queen of Thanet*, and another 7,000 by *Medway Queen*, and 16,000 by *Royal Sovereign*.

Three officers of *Medway Queen* whilst she was serving as a minesweeper during the Second World War. The photograph was taken aboard the steamer around 1940. *Medway Queen*, like most other pleasure steamers, was requisitioned at the outbreak of the Second World War. Her greatest hour was at Dunkirk when she carried around 7,000 troops from the beaches. She was such a long time reaching land on her last trip that she was listed as missing. *Medway Queen* then went to serve as a minesweeper based mostly on the River Tyne. She was reconditioned by Thorneycroft at Southampton at the end of hostilities and returned to service in 1947.

Harold Collard Stone and his daughter Margaret examining one of the lifeboats from *Royal Sovereign*, which was mined and sunk in the Bristol Channel in 1940. Captain Tommy Aldis was in command of *Royal Sovereign* when she was lost in the Bristol Channel. Thirty-two men were badly injured and his chief officer was sadly killed. The photograph was taken on Berrow Sands. Harold was a keen pleasure steamer historian and fan of the River Medway fleet. He had a special affinity with the motor ships. On the day when this photograph was taken, Harold collected wood from the wreckage and later made a model of *Royal Sovereign*. This model survives in the PSPS Collection and may now be the only part of this famous pleasure steamer to survive.

Chapter 3

Postwar Years

The paddle box of *Medway Queen* in 1963. The years after the Second World War witnessed the decline of the River Medway fleet.

With the cessation of hostilities in 1945, it would have been logical to have recommenced services as they were in 1939. Surely, the future of the Medway and Thames fleet was assured as passengers would once again flock to the seaside again for fun after suffering the ravages of war for six years. But, the combined Eagle & Queen Line fleet had shrunk to six steamers from a pre-war total of thirteen. The only remaining New Medway Company steamers were *Medway Queen*, *Queen of Kent* and *Queen of Thanet*. Limited services resumed in 1946 and a more normal service followed a year later. It was clear that things would be very different to the way they were in 1939.

By this time, the New Medway Steam Packet and General Steam Navigation steamers were regarded as one fleet, although the New Medway fleet kept their registration identity and the company also managed its busy shipyard at Rochester. Fleet replacements for the war losses *Queen of the Channel* and *Royal Sovereign* were ordered, and for some, it seemed as if the confidence of the pre-war years would continue. After derequisition in 1945, *Queen of Thanet* was reconditioned by the New Medway Company. She re-entered passenger service in 1946 after the completion of five months of restoration. Excursion trade recommenced on 'Victory Day', 8 June 1946, when *Queen of Thanet* sailed from the Medway towns to Southend and Margate. Routes at the time were severely restricted due to the remaining danger of mines and wrecks. These destinations were some of the few places that were deemed safe. Sadly, her pre-war popularity was never repeated, and after two short seasons, she left for further service on the South Coast. 1946 also saw *Queen of Kent* derequisitioned. With her rusty hull and noisy old engines, she was hardly a prime candidate for a costly refit. However, the New Medway Company had suffered significant losses, and although brand-new tonnage was to be built, a shortage of steel meant that there wouldn't be a fast solution to their needs. *Queen of Kent* therefore underwent a partial refit to fill the (albeit short) gap. With the entry into service of the resplendent new *Royal Sovereign*, *Queen of the Channel* as well as the smaller *Rochester Queen*, the aged and old-fashioned-looking *Queen of Thanet* and her sister *Queen of Kent* were laid up and sold as surplus to requirements. They had originally been laid up in the 1939 season after the entry into service of the new motor ships. They had ably filled the gap in the initial years after the end of the Second World War whilst replacement motor ships were constructed, but like in 1939, their demise was imminent when the motor ships entered service.

Medway Queen valiantly soldiered on alone as the last Thames and Medway paddle steamer. By the late 1950s, the situation was looking very bleak indeed. What was once a trickle of withdrawals of pleasure steamers was now becoming a torrent.

Finally, the inevitable happened and the bright new world of the 1960s saw *Medway Queen* finally withdrawn from service. But her withdrawal was met by a huge public outcry. A campaign was started to save the 'Heroine of Dunkirk', and after a number of false alarms and near disasters, she was towed to the Isle of Wight to take up a new life. It seemed that the last River Medway paddle steamer would at least find a new home in a world that was rapidly losing that once-common and hugely nostalgic vessel – the paddle steamer.

This rare photograph shows *Royal Daffodil* being reconditioned by Denny Bros at Dumbarton after wartime service. She was derequisitioned in January 1947 after carrying around 2.5 million servicemen and sailing almost 200,000 miles. Externally, she was rebuilt to look like a new pleasure steamer but her engine-room was different, as she had been damaged by a bomb at Dunkirk in 1940. The photograph shows her without her second funnel and with welding taking place. A year later, in 1948, *Royal Daffodil* was given a replacement engine to replace the war-damaged one. Her return journey to Rochester in February 1949 was undertaken in perfect conditions and she steamed from Gourock to Sheerness in just forty-five hours. On trials, she had reached a speed of 20.8 knots.

Queen of Thanet being refitted at Rochester. *Queen of Thanet* was the first Thames pleasure steamer to sail after the end of the Second World War. On Victory Day, she commenced her first postwar trip, beating *Royal Eagle* by nearly an hour, and was the first to arrive at Southend and Margate.

An aerial view of the postwar *Royal Sovereign*. She was the last link with the motor ships created by Sydney Shippick. She survived in the Bay of Naples as a much-transformed ferry until finally being broken up in December 2007.

Medway Queen at Rochester. *Medway Queen* appeared in the 1953 Coronation Spithead Fleet Review. Both *Royal Daffodil* and *Royal Sovereign* departed from London for Spithead with *Medway Queen* departing from Chatham. By Dover, there was thick fog and passengers aboard the two large steamers felt sorry for the poor old *Medway Queen* without radar, etc. When the two large pleasure steamers went into Southampton Water, they noticed the gallant paddle steamer from the River Medway, *Medway Queen*, happily taking bunkers at Hamble! She had beaten the weather and her two bigger and faster sisters!

An unusual view of *Medway Queen* cruising between Southend and Herne Bay around 1959. *Medway Queen* carried 66,000 passengers during that season.

Rochester Queen at sea around 1952. She had been built in 1944 as a military landing craft by the Stockton Construction Company. She was purchased in 1947 by the New Medway Steam Company for passenger service. She entered service in 1948 after being refitted by the company.

Rochester Queen was 147 feet in length. On her trials, *Rochester Queen* achieved a speed of 12 knots and could carry up to 425 passengers. She was initially placed on the Strood, Southend and Herne Bay run, calling at Sheerness Pier until 1954. For a while, around 1953, *Rochester Queen* was based at Clacton for local cruises.

SPECIAL ★ ★
AFTERNOON SAILINGS

On Sunday 11th, Monday 12th, Tuesday 13th, Wed. 14th, Thurs. 15th & Saturday 17th JUNE—ONLY

THE POPULAR TWIN-SCREW M.V.

"ROCHESTER QUEEN"

will sail from **SOUTHEND PIERHEAD**

(Weather and other circumstances permitting)

at **2.15** p.m.

to allow 2 HOURS ASHORE *at*

SHEERNESS-ON-SEA

Return Fare **4/6**   Child **2/3**

Tickets Admit to Pier

OUT			— TIMES —		HOME		
SOUTHEND	dep.	2.15 p.m.		SHEERNESS	dep.	5. 0 p.m.
SHEERNESS	arr.	3. 0 p.m.		SOUTHEND	arr.	5.45 p.m.

OPPORTUNITIES YOU SHOULD NOT MISS!

TEAS Etc. ON BOARD   —   FULLY LICENSED

Book at **EAGLE & QUEEN LINE OFFICES**
or **ON THE STEAMER**   P.T.O.

Handbill for special afternoon cruises by the *Rochester Queen* during the 1950 season. She had a dining saloon, refreshment bar and lounge.

The dining saloon of *Medway Queen* around 1950. This somewhat cosy saloon was very different to the vast dining saloons of the three large motor ships. *Medway Queen* had a framed portrait of Elizabeth II displayed in this saloon for many years.

During the 1955 and 1956 summer seasons, *Medway Queen* ran daily (except Fridays) from Strood and Chatham to Southend and Herne Bay. In 1957, due to the removal of the *Queen of the Channel* and *Crested Eagle* from the area, *Medway Queen* started running via Southend to Clacton and then went on a River Blackwater cruise on Tuesdays and Saturdays.

Medway Queen on the Medway during a winter refit. Throughout her entire career, *Medway Queen* had two long-standing masters. Her first was Captain Bob Hayman during the 1920s and 1930s and her second and final one was the popular Captain Leonard Horsham in the years after the Second World War. Hayman was a big and often smiling man who served her well in her early years. After the war, he became ill and died shortly after. He requested that his ashes be scattered in Gillingham Reach, so that his beloved steamer would pass over his ashes each day.

During 1959, *Medway Queen* undertook runs to Clacton on Wednesdays and Saturdays. On the other days (except Fridays) she cruised to Herne Bay. During the 1960 and 1961 seasons, *Medway Queen* continued with the same pattern of cruises as 1959.

QUEEN LINE of STEAMERS

DELIGHTFUL
SEA CRUISES
From STROOD & CHATHAM
Commencing WHIT-SAT. 28th MAY

DAILY
(excepting Fridays, Sat., 4th JUNE, Thur., 16th JUNE & Sun., 28th AUG.)
TO
SOUTHEND 7½ hours ashore **7'6**
HERNE BAY 4½ hours ashore **8'-**

From STROOD PIER   -   9.00 a.m.
SUN PIER, CHATHAM   -   9.20 a.m.
Return Times from HERNE BAY 5 p.m., SOUTHEND 6.40 p.m..

To **MARGATE**   (change at SOUTHEND)   **10/-**
Leave STROOD PIER 9.00 a.m. and SUN PIER, CHATHAM 9.20 a.m.

Commencing SAT., 18th JUNE
DAILY (excepting Fridays, SAT., 25th JUNE & SUN., 28th AUG.)
To **CLACTON**   (change at SOUTHEND)   **11/-**
Leave STROOD PIER 9.00 a.m. and SUN PIER, CHATHAM 9.20 a.m.

Period Return Fares to Southend 11/- ; Herne Bay 11/-
Margate 13/- and Clacton 17/-

CATERING, REFRESHMENTS & FULLY LICENSED BARS ON BOARD

CHILDREN 3 to 14 years HALF FARE

NOTE :   The Company regrets that there will be no services to SHEERNESS
until further notice owing to SHEERNESS PIER being closed.

Bicycles 3/- single.   Luggage allowed to Single and Period Passengers only.—
Passengers are carried only on the terms and conditions printed on the back of
this handbill.   No dogs allowed on board.
All Sailings are subject to weather and other circumstances permitting.

Tickets can be purchased in advance at SUN PIER CHATHAM or from
NEW MEDWAY STEAM PACKET Co., Ltd.
Head Office: 366 HIGH STREET, ROCHESTER   Phone CHATHAM 2204-5
G.S.1—1955   R.A.P. Ltd.

A handbill familiar to anybody that travelled aboard the *Medway Queen* from the Medway towns during the 1950s. It outlines her then-regular service to Southend, Herne Bay and Clacton. You could spend 7.5 hours ashore at Southend or 4.5 hours ashore at Herne Bay with the option of changing steamer at Southend for Margate and Clacton. Sheerness was no longer used for passenger calls after the 1954 season. Both *Medway Queen* and *Rochester Queen* had to change their schedules due to this.

The immaculate engine-room of the *Medway Queen*. She had a compound diagonal engine. In 1938, she was converted to burn oil. Her bunkerage was small, which necessitated refuelling from the barge *Mudd* on alternate days.

Officers 'towing the lead' aboard *Medway Queen* whilst on a cruise to Clacton. Regular passengers particularly looked forward to when, owing to a low tide, the return trip by *Medway Queen* from Clacton was by the long route around the Gunfleet Sands. This added up to 17 miles and 1.5 hours to the journey. Captain Leonard Horsham once jokingly remarked that they knew the dates before he did!

Medway Queen on her way to Southend and Herne Bay passing Rochester. Leonard Horsham was her last master. This well-loved River Medway figure died in 1969. *Medway Queen* had a big reputation for having a strong personality and was seen as a very friendly steamer. Most weekends, paddle steamer fans would meet up to enjoy each other's company and to mix with the friendly customer-focused crew.

Medway Queen dressed with flags alongside Margate Jetty during a rare charter cruise in 1962. On top of her funnel you can see the apparatus fitted at the time, as she had no main mast and navigation lamps needed to be fitted.

Medway Queen approaching Herne Bay Pier towards the end of her career. *Medway Queen* spent two-thirds of her time in shallow water and had a draft of just 5.5 feet. She carried 828 passengers on the Southend run.

Medway Queen alongside Strood Pier. This was partially rebuilt in pre-stressed concrete in 1960.

Medway Queen approaching Southend Pier during a quiet day in the 1950s. Her record for maintaining services was second to none and she always ran, even if she had only twenty passengers aboard.

Leonard Horsham on the bridge of *Medway Queen*. She retained an open bridge until the end.

Medway Queen was famous for her appearance; the reason for this can be clearly seen from this photograph showing gleaming brass and woodwork.

The engine of *Medway Queen* photographed from below the control platform. Even during her final days of service, *Medway Queen* was still in the limelight. On 30 May 1963, she was chartered at Herne Bay to be used during the filming of the film *French Dressing*. The film was set in a dreary, fictional seaside town named Gormleigh-on-Sea and featured Roy Kinnear and James Booth. The following day, *Medway Queen* was chartered as relief for *Royal Sovereign* for the Thames barge race and because of this made a very rare call at Gravesend. A few days later, on 17 June, her Clacton trip was replaced by a special sailing from Strood to view the Medway barge race.

A rare shot of the interior of the engine-room of the *Medway Queen* adjacent to the engine. *Medway Queen* was economical on fuel and only used around 3.5-4 tons a day. The chief engineer aboard *Medway Queen* for many years was Norman Taylorson.

No. 544	No. 544
THE NEW MEDWAY STEAM PACKET CO., LTD.	THE NEW MEDWAY STEAM PACKET CO., LTD.
CHARTER OF P.S. "MEDWAY QUEEN"	CHARTER OF P.S. "MEDWAY QUEEN"
BY	BY
D. ROSE LTD.,	D. ROSE LTD.,
THURSDAY, 16TH JUNE, 1955	THURSDAY, 16TH JUNE, 1955
Leave TOWER PIER at 8.20 a.m.	Leave CLACTON 4.15 p.m.
TO	TO
CLACTON	TOWER PIER
This portion to be given up on landing at Clacton	This portion to be given up on landing at Tower Pier
FOR CONDITIONS OF CARRIAGE SEE OVER	FOR CONDITIONS OF CARRIAGE SEE OVER

Ticket for Don Rose's charter of *Medway Queen* from Tower Pier to Clacton on 16 June 1955. It was arranged as a thank-you to his customers after wartime rationing and disruption. Don later became involved with ventures such as *Consul's* brief visit to the Thames and Medway in 1963 and the short-lived career of *Queen of the South* a few years after that.

Queen Line of Steamers

1959

SPECIAL NOTICE

As a result of the closing of the pier head at Sun Pier, Chatham, our steamer "Medway Queen" **will now operate from Strood Pier only.** To enable intending passengers to get to Strood in time, the steamer's departure from Strood will now be 9.15 a.m.

On weekdays frequent 'bus services from Military Road, Chatham pass Canal Road, Strood (first stop on the Strood side of Rochester Bridge).

Strood station is adjacent to Strood Pier and on *Sundays* a train departs from:—

Sittingbourne		8.39
Rainham		8.48
Gillingham	*8.24 and	8.55
Chatham	*8.28	8.59
Rochester	*8.30	9.1
arrive Strood	*8.33	9.5

*Sunday, 14th June

first train shown only—depart 14 minutes later

1959 saw the closure of Sun Pier at Chatham. It had been closed due to its dangerous condition. Chatham Council estimated that it would cost £30,000 to restore and therefore hoped to demolish it. This had serious consequences for *Medway Queen*, as she then had to pick up her passengers from Strood Pier only. Although only a short bus ride away, the closure of Sun Pier meant that *Medway Queen* witnessed a drop in passenger numbers. It was also stated at the time that residents of Chatham and Rochester couldn't find Strood Pier. The closure of Sun Pier must have put one of the final nails in the coffin of *Medway Queen*.

A view of the bow and bridge of the *Medway Queen* whilst alongside Southend Pier in July 1962. You can see the New Medway Steam Packet house flag emblem attached to the funnel that stayed with *Medway Queen* throughout her long career.

Medway Queen alongside Strood Pier in 1962 looking towards Rochester Bridge. She had spent her entire career of 39 years (minus her wartime service) passing this landscape.

Medway Queen passing Rochester during the late 1950s. Captain Leonard Horsham, along with the regular team of Chief Engineer Bill Ruthven, Chief Steward Tom Gibb and Tom, the Bosun, made a trip aboard *Medway Queen* a good and happy one.

THAMES & MEDWAY NAVIGATION CO.

COMMENCING 4th AUGUST, 1959
SALOON SHIP
"ANZIO"

LEAVES SOUTHEND PIER DAILY

at **8.30** a.m.

for

CHATHAM

allowing **8 HOURS ASHORE**

ARRIVING AT GILLINGHAM PIER AT 10 a.m.
LEAVING GILLINGHAM AT 6 p.m.
ARRIVING BACK AT SOUTHEND ABOUT 8 p.m.

RETURN FARE—7/6
Children Half-price

FULLY LICENSED AND REFRESHMENT BARS

This Vessel can be hired for Private Parties
All sailings are subject to weather and other circumstances permitting.

Booking Office at Pier Entrance and Pier Head

Printed by Osier-Ritter & Son, 380 London Road, Hadleigh, Essex.

Handbill for the first season of *Anzio* on the Medway and Thames in 1959. She commenced service on 4 August and provided a ferry service between Southend and Sheerness and onwards to Gillingham. Sadly, *Anzio* was unable to take up services as advertised, as the authorities said that she needed to have a pilot onboard. She therefore began regular service to the Medway on 14 May 1960.

A view of the *Medway Queen*'s promenade
deck at Rochester around 1962. Although the
New Medway and General Steam Navigation
vessels were one fleet after the war, they each
kept their own separate identities.

A view aboard *Medway Queen* during her
last month of service in September 1963. Her
appearance had changed very little on the River
Medway during almost forty years of service.

Newspaper cutting showing BBC commentator John Richardson aboard *Medway Queen* testing the radio-telephone link with the pier. He is watched by Captain Leonard Horsham and programme producer Bill Duncalf. *Medway Queen* became quite a media celebrity in her final years. On 5 May 1961, Southern TV made a short film on the New Medway Company including footage of *Medway Queen*.`

Medway Queen on the gridiron at Rochester in 1961. By this time, her days were numbered and she was facing withdrawal. At the time of her withdrawal, the National Maritime Museum took a full pictorial record of *Medway Queen* – the last New Medway Steam Packet paddle steamer.

William Peake with his son Alan. At the end of 1937, Bill Peake was appointed Engineer Superintendent of the New Medway Steam Packet Company and had responsibility for all engineering aspects of the fleet. In 1951, Bill he was appointed a director of the New Medway Company, which continued to operate the paddle steamer *Medway Queen*. Bill took over as the Managing Director on the retirement of Mr O'Keefe, who, in turn, had taken over from Captain Shippick, who reformed the company after the First World War. Bill Peake died whilst in post as the Managing Director of the New Medway Steam Packet Company on 1 November 1962. His son Alan has kept up the tradition by becoming a director of *Kingswear Castle*.

Engine-room control platform aboard *Medway Queen* in 1962. During August 1962, it was rumoured that *Medway Queen* was to be withdrawn from service and an article in *The Guardian* stated that she was to be sold to the Paddle Steamer Preservation Society. The rumour was denied by her owners.

When the New Medway Steam Packet Company offered *Medway Queen* for sale in 1963, they stipulated in the advertisement that she would have to be purchased 'with guaranteed demolition'. After her last trip, it was intended that she would be stripped ready for scrapping.

Medway Queen said her farewell to regular passenger service on the River Medway on Sunday 8 September 1963. Over 600 passengers left Strood Pier to sail for the last time down the River Medway on a New Medway Steam Packet steamer. Around 400 were dropped off at Southend with around 350 more joining (many being journalists). The final scene at Herne Bay saw the Chairman of the Council, Mrs Gwendoline Fortune go aboard *Medway Queen* to make a speech touched with emotion, hoping that the PSPS would be able to help save her. They bade farewell from the pier over the loudspeakers: 'Bon voyage Captain Horsham, bon voyage *Medway Queen*.' As the ship departed from Southend Pier for the last time, crowds were entertained by a recording made by members of the pier concert party singing 'Now is the Hour'. With streamers, rockets and cheers from those waiting on the pier, *Medway Queen* headed home for the last time with her 700 passengers. She arrived back at Strood Pier at 9.25 p.m. with many solemn-looking and sad passengers.

Frank Marshall, the secretary of the New Medway Company, announced to members of the Paddle Steamer Preservation Society in 1963 that *Medway Queen* had ceased to provide them with any 'jam' since 1959. Competition had come from the motor car and the better weather abroad. She was seen as a second interest to the company after ship repairing, which was increasing. The meeting was very favourable towards preserving *Medway Queen* until it was learnt that her sale by the New Medway Steam Packet Company was subject to her not being placed in commercial service. Despite this, an independent *Medway Queen* Trust was formed to preserve the last River Medway paddle steamer and to place her at a suitable mooring. The treasurer was Don Rose, who had earlier chartered *Consul*.

Medway Queen approaching Clacton Pier for the final time in 1963. In its heyday, this popular and bustling pier included many entertainments, such as the Blue Lagoon Ballroom, Ocean Theatre and a large open-air swimming pool. It was thought at the time that no paddle steamer would ever visit the pier again. The steamer berthing arm is still used by *Waverley* and *Balmoral*, although it has deteriorated as a calling point since its heyday.

Medway Queen was moved by tugs on January 29 1964 from Sheerness to Nelson Dock at Rotherhithe under the command of Captain Horsham. She is shown here at the dock. One plan for *Medway Queen* after her retirement was for her to be steamed around the coast of the UK in connection with the National Trust's Coastal Preservation Scheme. It was envisaged that she would be moved with a skeleton crew with a civic reception held at each port. It was further envisaged that, in 1965, she would be refitted to get a Class 5 certificate to act as the flagship of the flotilla to return to Dunkirk to commemorate the twenty-fifth anniversary of the evacuation.

MEDWAY QUEEN TRUST
═ NATIONAL APPEAL ═

SAVE THE MEDWAY QUEEN

The Famous Dunkirk Paddle Steamer

£8,000 URGENTLY NEEDED

SEND YOUR DONATION NOW

to The Appeal Treasurer, 500 Old Kent Road, London, S.E.1

Lanes, Printers, Broadstairs & Herne Bay

The Paddle Steamer Preservation Society decided to hold a public meeting after *Medway Queen* was withdrawn. It was held on 18 October 1963 at the Baltic Exchange in London. Over 100 people attended, and it was decided to hold a public appeal to raise £8,000 to manage the ship under the auspices of the '*Medway Queen* Trust'. The New Medway Company agreed to allow the Trust until the end of 1963 to raise the money. This poster was used in the fundraising campaign.

Medway Queen in the East India Dock in June 1964. This shows the promenade deck looking towards the stern.

Medway Queen in the East India Dock in June 1964 waiting to hear her fate. *Medway Queen* was sold on 13 August 1965 for the sum of £5,000 for scrapping by Belgium shipbreakers. Miraculously, the veteran of the River Medway was saved at the eleventh hour by three Isle of Wight businessmen. Backed by the PSPS and the *Medway Queen* Trust, they purchased her for £6,000.

Medway Queen on the River Medina on 28 September 1965. The photograph shows her just after her tug from London had handed her over to local boats *Match* and *Seaclose*. She had just been turned ready to take her up the Medina stern first and into the breach of the wall between the river and the millpond.

Medway Queen was now at her new home – a long way away from the River Medway. Everyone was hopeful that she would face a bright future. Sadly, this wasn't to be and she returned to the Medway in the 1980s. Thanks to the Heritage Lottery Fund, a restoration project started in 2009.

Two 1924-built Medway paddle steamers meet for the first time on the ... Isle of Wight! *Kingswear Castle* is moored on the River Medina close to *Medway Queen* in August 1967. Both steamers look smart, but within a few short years, their futures became more uncertain. They were reunited on the River Medway some twenty years later when the derelict *Medway Queen* was towed back to her native river to be met by the then-operational *Kingswear Castle*.

Royal Eagle moored in Whitewall Creek waiting to be scrapped during the early 1950s. The Medway was sometimes used as a pleasure steamer graveyard for steamers awaiting demolition elsewhere. Another famous paddle steamer that awaited this fate on the Medway was *Golden Eagle*.

The post-war *Queen of the Channel* laid up at Rochester on the River Medway on 16 April 1966. Many people will remember the GSN motor ships that were frequently laid up each winter on the River Medway close to Rochester.

Three well-known figures from the heyday of the New Medway Steam Packet Company. *Queen of the South* (ex-*Jeanie Deans*) had a formidable trio of officers aboard her during the 1967 season. Her master was Captain George Fowle (centre) aged sixty-eight, who forty years earlier was the first officer on the New Medway Steam Packet's *Queen of Thanet*. The Chief Officer was sixty-nine-year-old Tommy Aldis, DSC, (left) who had been the master of *Queen of Thanet* and right-hand man to Captain Sydney Shippick during the 1920s and 1930s, when the New Medway Steam Packet Company grew in importance. He had commanded the first *Royal Sovereign* as she evacuated children from London to the East Coast resorts in September 1939. In 1940, he commanded her at Dunkirk, La Panne, Cherbourg and St Malo, bringing back over 20,000 men. Later, when *Royal Sovereign* was struck by a mine, Captain Aldis was seriously injured. He was awarded the Distinguished Service Cross and Lloyd's Medal and was mentioned in despatches. In 1941, he joined *City of Rochester*, which was later sunk by a landmine in the River Medway. The chief engineer was seventy-year-old Tommy Williams (right) who had been with his two colleagues on *Queen of Thanet* some thirty-five years earlier.

The Paddle Steamer Preservation Society was formed in 1959 to help preserve paddle steamers around the UK. The 1960s saw a rapid decline of the paddle steamer fleets around the UK. There were also brave attempts to operate redundant steamers such as *Queen of the South* (ex-*Jeanie Deans*). She is shown here tied up against buoys at Chatham on Saturday 3 September 1966.

The postwar *Queen of the Channel* lying on the Medway at Rochester on 25 February 1968. Her funnel is painted with the 'K' of her new owners (John P. Katsoulakos). Just over a week later, she sailed from Sheerness to Greece to embark on a second career as *Oia*. It seemed for many that the River Medway would never see a pleasure steamer again.

Chapter 4

The 1970s Onwards

Kingswear Castle cruising in the Thames Estuary around 1987 with Southend in the distance. The 1980s witnessed a splendid revival of paddle steamer cruises on the River Medway with the arrival of *Kingswear Castle*, supported by *Waverley* and *Balmoral*.

September 1963 signalled the end of regular River Medway steamer services when the last New Medway Steam Packet Company paddle steamer, *Medway Queen*, was withdrawn. This withdrawal was viewed within a world where just about every pleasure steamer was ending its days. Was there a future for further passenger steamers on the River Medway and would the Medway towns ever see a paddle steamer again?

The mid-1960s witnessed a few brief visits by pleasure steamers, but it wasn't until 1971 that a paddle steamer returned again. This time it was a visitor from the River Dart in Devon – *Kingswear Castle*. *Kingswear Castle* then underwent a lengthy restoration of over ten years, until the early 1980s, when she was steamed again. She eventually re-entered full passenger service in 1985. Since that date, she has regularly plied the waters of the Medway, becoming a firm favourite of tourists and locals alike. Two other regular visitors to the River Medway, *Waverley* and *Balmoral*, have visited the area since the early 1980s, recreating the classic cruises from the glory days of the New Medway Steam Packet era. Both steamers have also steamed together with *Kingswear Castle* each year in the 'Grand Parade of Steam' on the River Medway, making the river the only place in the UK where you can see two operational paddle steamers parading together.

There have also been many other more infrequent visitors that have been based or have offered cruises to the River Medway. These have included *Clyde*, which offered regular cruises to Southend during the late 1980s, and *Princess Pocahontas* has also visited from its home in Gravesend.

The last survivor of the New Medway Steam Packet Company – *Medway Queen* – also returned to her old home in the 1980s. A preservation society was formed to save and preserve her, and in 2009, a large-scale restoration project was started.

The River Medway can therefore be seen as a river where the great tradition of pleasure steamer cruising is alive and indeed thriving, with its regular paddle steamer *Kingswear Castle*, along with *Waverley* and *Balmoral*. The future is therefore bright for those who enjoy a carefree and relaxing cruise on a paddle steamer – a tradition that goes back almost two centuries. Long may this River Medway tradition continue.

Kingswear Castle during her heyday on the River Dart in the 1920s. She is seen passing the two castles at the mouth of the river. Despite there being a glut of redundant paddle steamers during the early to mid-1960s, *Kingswear Castle* became the lucky steamer that was saved by the Paddle Steamer Preservation Society. This was due to her size, heritage and huge suitability for restoration and operation.

Kingswear Castle at Medway Bridge Marina on 25 February 1973. Her restoration was very slow and it took another ten years before *Kingswear Castle* was operational again. *Kingswear Castle* was the first steamer saved for preservation and operation by the Paddle Steamer Preservation Society.

A section looking towards the bow of *Kingswear Castle* whilst it was being restored on the Medway around the mid-1970s. The deck is being replaced and you can see down into the saloon below. You can appreciate from this photograph how much the steamer had deteriorated since her time on the River Dart.

On 20 November 1982, Rochester City Council presented the *Kingswear Castle* team with a cheque towards the 'Deck Timber Fund'. Nick Knight (left) is seen accepting the cheque from Councillor Rodberg at the Medway Bridge Marina. Volunteers involved with the restoration look on.

Kingswear Castle's exposed paddle wheel viewed whilst undergoing restoration. As well as volunteers, many local businesses contributed to the restoration by donating materials, time and equipment.

Kingswear Castle's engine dates back to 1904, some twenty years before the ship was built. It was transferred from the previous paddle steamer of the same name. The engine was restored on the River Medway ready for the first steaming in November 1983.

Prior to entry into service, *Kingswear Castle* had several people acting as master. They included John Megoran (centre) and Dan Macmillan (right).

Captain John Megoran alongside *Kingswear Castle* around 1985. John Megoran was appointed Master and General Manager of *Kingswear Castle* in 1985 and since that time has influenced every aspect of the steamer's operation. *Kingswear Castle* has gained a high reputation as the River Medway's operational paddle steamer. John grew up in Weymouth and was a keen passenger aboard the famous Cosens steamer fleet during their final years of operation. Running a paddle steamer on the Medway is a year-round activity.

Kingswear Castle on her first run after restoration during her steaming trials on the River Medway on 4 November 1983. It was the culmination of over ten years of restoration by a dedicated band of volunteers.

A view showing *Kingswear Castle*'s almost empty decks in 1984. At the time, she was only able to carry a small number of passengers, until a full passenger certificate was issued. Over the following few years, her deck came to life again with seating, awnings, new Bridge wings and doors.

It had taken almost twenty years for *Kingswear Castle* to be restored after her withdrawal from service on the River Dart. She was purchased by the Paddle Steamer Preservation Society for £600.

Since her restoration on the River Medway and re-entry into full passenger service, *Kingswear Castle* has had to develop to face an ever-increasing range of passenger safety and other regulations. She has also had a new boiler and a new hull below the waterline to ensure that she gives pleasure on the Medway for many years to come.

Kingswear Castle met with *Waverley* for the first time in the River Medway in 1984. Crew members shown here are Ken Blacklock (Chief Engineer *Waverley*), Chris Jones (Chief Engineer *Kingswear Castle*), Captain David Neill (Captain of *Waverley*), and Roger Tuft (*Kingswear Castle* crew).

Waverley passes *Kingswear Castle* at their first meeting on the Medway on 9 September 1984.

The smart aft saloon of *Kingswear Castle* around 1986. It shows the old-fashioned ambience of the 1924 steamer, which was restored so well by volunteers. This saloon was a bar for the first few seasons of service. Lighter refreshments were served in the other saloon towards the bow until both were combined around 1990.

Every October, *Kingswear Castle* meets *Waverley* on the River Medway for the 'Grand Parade of Steam'. The UK's two most famous paddle steamers celebrate the work of the Paddle Steamer Preservation Society during this enjoyable event.

Kingswear Castle approaching Strood Pier on 26 June 1999.

Handbill advertising a 'Medway Mystery Cruise' aboard *Kingswear Castle* for her final sailing of the 1986 season on 5 October.

Kingswear Castle has travelled far away from her Chatham base during a quarter century of operation on the River Medway. This has included many trips to London (as far as Putney) as well as Herne Bay and Whitstable.

Kingswear Castle departs on a Christmas cruise from Thunderbolt Pier at Chatham Historic Dockyard around 1990.

Timetable for *Kingswear Castle* for the 1987 season. At this time, she offered regular short cruises from Southend Pier. *Kingswear Castle* was lucky in that the completion of her restoration coincided with the opening of Chatham Historic Dockyard as a major tourist attraction. She has also undertaken cruises as part of the 'Dickens Festival' since the 1980s.

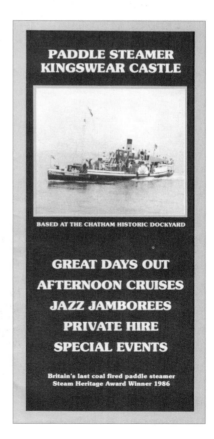

PADDLE STEAMER
KINGSWEAR CASTLE

BASED AT THE CHATHAM HISTORIC DOCKYARD

GREAT DAYS OUT
AFTERNOON CRUISES
JAZZ JAMBOREES
PRIVATE HIRE
SPECIAL EVENTS

Britain's last coal fired paddle steamer
Steam Heritage Award Winner 1986

Kingswear Castle departing from Strood Pier on 14 September 1989. *Kingswear Castle* regularly used Strood Pier from her first sailings on the Medway in 1984 until its closure around twenty years later.

Timetable leaflet for *Kingswear Castle* for the 1988 season. By this time, *Kingswear Castle* had settled into a regular pattern of sailings, combining short afternoon trips with evening jazz cruises and longer days out to places such as London, Whitstable and Southend. Regular events included participation in the annual 'Dickens Festival' and 'Gravesham Edwardian Fair'.

Kingswear Castle has undertaken cruises up as far as New Hythe during her career on the Medway. She is seen here around 1987 amidst the typical 'upriver' scene of high reeds, twisting bends and paper mills.

In 1986, *Kingswear Castle* won the National Steam Heritage Award and, in 1995, won first prize in the Scania Transport Trust Awards. In 1999, she was included on the National Historic Ships Committee Core Collection list of ships of 'Pre-eminent National Significance'. Since being on the Medway, she has also welcomed famous people such as HRH Prince Edward, Margaret Thatcher, Sir Harry Secombe and Pierce Brosnan aboard.

Leaflet advertising discount cruises aboard *Kingswear Castle* for those travelling to join the steamer by train at Strood Pier during the 1988 summer season. Generous discounts were given on regular afternoon cruises as well as longer day cruises on the steamer. The location of Strood Pier was close to the railway station, giving passengers the convenience of just a few minutes walk to the steamer.

Kingswear Castle alongside Chatham Sun Pier during the late 1980s. A landing pontoon had been built for passenger traffic at the time after many years of neglect. *Kingswear Castle* continued to use Sun Pier for several years before it closed again.

Kingswear Castle and *Waverley* taking part in the annual autumn 'Parade of Steam' on the River Medway off St Mary's Island on 21 September 1985. Just fifteen years earlier, both paddle steamers had faced uncertain futures. The annual meeting of the two paddle steamers on the River Medway celebrates the tremendous work undertaken to preserve and operate these historic steamers.

Medway Queen was moored alongside the old pumping station at the dockyard for several years during the mid-1980s. She arrived at this location in 1985, and during the next couple of years, her appearance deteriorated. She became almost submerged at high tide. At this time, several key figures entered the scene and, with a growing group of individuals, publicised the plight of the last New Medway Steam Packet Company paddle steamer – *Medway Queen*.

The Invicta Line's *Clyde* was a familiar sight on the River Medway during the late 1980s, when she conveyed passengers between Strood and Southend.

Waverley transfers passengers to *Kingswear Castle* in the River Medway around 1995. *Kingswear Castle* now carries a maximum of 235 passengers on her cruises. On her native River Dart, she could carry an astonishing 504!

Queen Mary exiting No. 3 Dry Dock at Chatham Historic Dockyard in July 1988. *Queen Mary* had spent her career on the Firth of Clyde until retirement. She was refitted at Chatham for further use as a floating bar and restaurant on the Embankment in London. She stayed in this role until 2009. *Queen Mary*'s old funnel was taken to the Doust Yard at Rochester for disposal during the refit.

An unusual sight as the paddle tug *John H. Amos* is lifted out of the River Medway in March 2008. At the time, she had to be moved from her berth beside the Historic Dockyard as development was about to take place in that area. *John H. Amos* arrived several years earlier on the River Medway for restoration.

Waverley passing Thunderbolt Pier and the covered slips at Chatham Historic Dockyard around 1990. *Waverley*, as well as her consort *Balmoral*, made several calls at Chatham Historic Dockyard during the 1990s to disembark passengers. This has now ceased due to the boat moorings. *Waverley* also made calls at the Bulls Nose entrance to the Chatham Dockyard.

Balmoral on the River Medway close to Chatham Dockyard during October 1987. This was *Balmoral*'s first visit to the Medway whilst she was deputising for *Waverley*, which was experiencing operational problems. *Balmoral* has become a frequent and popular visitor to the Medway since that time.

Balmoral has made visits to the River Medway since the mid-1980s when she re-entered service as a consort to *Waverley*. A popular rendezvous before 2000 was when *Balmoral* or *Waverley* met up with *Kingswear Castle* to exchange passengers. This image shows *Balmoral* alongside *Kingswear Castle* at Thunderbolt Pier. Stores are about to be transferred from *Kingswear Castle*.

Waverley passing Anchor Wharf at Chatham Historic Dockyard around 1990. *Waverley* has been a regular visitor to the River Medway for over thirty years. She entered service on the Clyde in 1947 and was sold for just £1 to the Paddle Steamer Preservation Society in 1974. She has since travelled to all parts of the UK. She makes regular visits to the Thames and Medway in September and October each year and regularly meets up with the charming *Kingswear Castle*.

Kingswear Castle is now the River Medway's only regular operational paddle steamer and the River Medway is the only river in the UK with a traditional paddle steamer in regular service for the entire summer period. *Kingswear Castle* is also Britain's last coal-fired paddle steamer and carries on the great tradition of paddle steamer cruises on the River Medway. By taking a cruise on her, you're able to sample the nostalgic transport of another age as well as being able to appreciate the fine maritime heritage that the River Medway can offer. As she cruises the river each summer, she keeps alive the great tradition of River Medway paddle steamers! www.kingswearcastle.co.uk